Handbook of Regression Modeling in People Analytics

Handbook of Regression Modeling in People Analytics

With Examples in R and Python

Keith McNulty

CRC Press

Taylor & Francis Group

Boca Raton   London   New York

CRC Press is an imprint of the
Taylor & Francis Group, an **informa** business

First edition published 2021

by CRC Press
6000 Broken Sound Parkway NW, Suite 300, Boca Raton, FL 33487-2742

and by CRC Press
2 Park Square, Milton Park, Abingdon, Oxon, OX14 4RN

© 2021 Keith McNulty

CRC Press is an imprint of Taylor & Francis Group, LLC

ISBN: 9781032041742 (hbk)
ISBN: 9781032046631 (pbk)
ISBN: 9781003194156 (ebk)

DOI: 10.1201/9781003194156

Typeset in Latin Modern font
by KnowledgeWorks Global Ltd.

Contents

Foreword by Alexis Fink   xiii

Introduction   xv

1 The Importance of Regression in People Analytics   1

  1.1 Why is regression modeling so important in people analytics?   2

  1.2 What do we mean by 'modeling'?   3

    1.2.1 The theory of inferential modeling   3

    1.2.2 The process of inferential modeling   5

  1.3 The structure, system and organization of this book   6

2 The Basics of the R Programming Language   9

  2.1 What is R?   10

  2.2 How to start using R   10

  2.3 Data in R   11

    2.3.1 Data types   13

    2.3.2 Homogeneous data structures   14

    2.3.3 Heterogeneous data structures   16

  2.4 Working with dataframes   18

    2.4.1 Loading and tidying data in dataframes   18

    2.4.2 Manipulating dataframes   22

  2.5 Functions, packages and libraries   24

    2.5.1 Using functions   24

    2.5.2 Help with functions   25

    2.5.3 Writing your own functions   26

DOI: 10.1201/9781003194156-0

2.5.4   Installing packages   26

2.5.5   Using packages   27

2.5.6   The pipe operator   28

2.6   Errors, warnings and messages   29

2.7   Plotting and graphing   31

2.7.1   Plotting in base R   31

2.7.2   Specialist plotting and graphing packages   33

2.8   Documenting your work using R Markdown   34

2.9   Learning exercises   37

2.9.1   Discussion questions   37

2.9.2   Data exercises   38

3   Statistics Foundations   39

3.1   Elementary descriptive statistics of populations and samples   40

3.1.1   Mean, variance and standard deviation   40

3.1.2   Covariance and correlation   43

3.2   Distribution of random variables   46

3.2.1   Sampling of random variables   46

3.2.2   Standard errors, the t-distribution and confidence inter-
        vals   47

3.3   Hypothesis testing   49

3.3.1   Testing for a difference in means (Welch's t-test) . . .   51

3.3.2   Testing for a non-zero correlation between two variables
        (t-test for correlation)   54

3.3.3   Testing for a difference in frequency distribution be-
        tween different categories in a data set (Chi-square test)   56

3.4   Foundational statistics in Python   58

3.5   Learning exercises   62

3.5.1   Discussion questions   62

3.5.2   Data exercises   63

4 Linear Regression for Continuous Outcomes **65**

4.1 When to use it 65

    4.1.1 Origins and intuition of linear regression 65

    4.1.2 Use cases for linear regression 66

    4.1.3 Walkthrough cxample 67

4.2 Simple linear regression 69

    4.2.1 Linear relationship between a single input and an outcome 70

    4.2.2 Minimising the error 70

    4.2.3 Determining the best fit 73

    4.2.4 Measuring the fit of the model 74

4.3 Multiple linear regression 76

    4.3.1 Running a multiple linear regression model and interpreting its coefficients 76

    4.3.2 Coefficient confidence 77

    4.3.3 Model 'goodness-of-fit' 78

    4.3.4 Making predictions from your model 81

4.4 Managing inputs in linear regression 82

    4.4.1 Relevance of input variables 83

    4.4.2 Sparseness ('missingness') of data 83

    4.4.3 Transforming categorical inputs to dummy variables . 84

4.5 Testing your model assumptions 86

    4.5.1 Assumption of linearity and additivity 86

    4.5.2 Assumption of constant error variance 88

    4.5.3 Assumption of normally distributed errors 89

    4.5.4 Avoiding high collinearity and multicollinearity between input variables 90

4.6 Extending multiple linear regression 93

    4.6.1 Interactions between input variables 93

    4.6.2 Quadratic and higher-order polynomial terms 96

4.7 Learning exercises 97

    4.7.1 Discussion questions 97

    4.7.2   Data exercises 97

5 Binomial Logistic Regression for Binary Outcomes   101

  5.1 When to use it 102

     5.1.1   Origins and intuition of binomial logistic regression . . 102

     5.1.2   Use cases for binomial logistic regression 103

     5.1.3   Walkthrough example 104

  5.2 Modeling probabilistic outcomes using a logistic function . . 106

     5.2.1   Deriving the concept of log odds 107

     5.2.2   Modeling the log odds and interpreting the coefficients 109

     5.2.3   Odds versus probability 110

  5.3 Running a multivariate binomial logistic regression model . . 112

     5.3.1   Running and interpreting a multivariate binomial logis-
           tic regression model 113

     5.3.2   Understanding the fit and goodness-of-fit of a binomial
           logistic regression model 116

     5.3.3   Model parsimony 120

  5.4 Other considerations in binomial logistic regression 122

  5.5 Learning exercises 124

     5.5.1   Discussion questions 124

     5.5.2   Data exercises 124

**6 Multinomial Logistic Regression for Nominal Category Out-
comes   127**

  6.1 When to use it 127

     6.1.1   Intuition for multinomial logistic regression 127

     6.1.2   Use cases for multinomial logistic regression 128

     6.1.3   Walkthrough example 128

  6.2 Running stratified binomial models 131

     6.2.1   Modeling the choice of Product A versus other products 131

     6.2.2   Modeling other choices 133

  6.3 Running a multinomial regression model 133

     6.3.1   Defining a reference level and running the model . . . 134

	6.3.2	Interpreting the model	136
	6.3.3	Changing the reference	137
6.4		Model simplification, fit and goodness-of-fit for multinomial logistic regression models	138
	6.4.1	Gradual safe elimination of variables	138
	6.4.2	Model fit and goodness-of-fit	139
6.5		Learning exercises	140
	6.5.1	Discussion questions	140
	6.5.2	Data exercises	141

7 Proportional Odds Logistic Regression for Ordered Category Outcomes **143**

7.1		When to use it	143
	7.1.1	Intuition for proportional odds logistic regression . . .	143
	7.1.2	Use cases for proportional odds logistic regression . . .	145
	7.1.3	Walkthrough example	145
7.2		Modeling ordinal outcomes under the assumption of proportional odds	148
	7.2.1	Using a latent continuous outcome variable to derive a proportional odds model	148
	7.2.2	Running a proportional odds logistic regression model	150
	7.2.3	Calculating the likelihood of an observation being in a specific ordinal category	153
	7.2.4	Model diagnostics	154
7.3		Testing the proportional odds assumption	155
	7.3.1	Sighting the coefficients of stratified binomial models .	156
	7.3.2	The Brant-Wald test	157
	7.3.3	Alternatives to proportional odds models	158
7.4		Learning exercises	159
	7.4.1	Discussion questions	159
	7.4.2	Data exercises	160

8 Modeling Explicit and Latent Hierarchy in Data **163**

8.1		Mixed models for explicit hierarchy in data	164

8.1.1   Fixed and random effects   164

8.1.2   Running a mixed model   165

8.2   Structural equation models for latent hierarchy in data . . .   170

8.2.1   Running and assessing the measurement model   173

8.2.2   Running and interpreting the structural model   180

8.3   Learning exercises   185

8.3.1   Discussion questions   185

8.3.2   Data exercises   185

9   Survival Analysis for Modeling Singular Events Over Time 187

9.1   Tracking and illustrating survival rates over the study period   189

9.2   Cox proportional hazard regression models   193

9.2.1   Running a Cox proportional hazard regression model .   194

9.2.2   Checking the proportional hazard assumption   196

9.3   Frailty models   197

9.4   Learning exercises   200

9.4.1   Discussion questions   200

9.4.2   Data exercises   201

10 Alternative Technical Approaches in R and Python   203

10.1 'Tidier' modeling approaches in R   204

10.1.1 The broom package   204

10.1.2 The parsnip package   208

10.2 Inferential statistical modeling in Python   209

10.2.1 Ordinary Least Squares (OLS) linear regression   209

10.2.2 Binomial logistic regression   211

10.2.3 Multinomial logistic regression   212

10.2.4 Structural equation models   213

10.2.5 Survival analysis   215

10.2.6 Other model variants   218

11 Power Analysis to Estimate Required Sample Sizes for Modeling   221

11.1 Errors, effect sizes and statistical power   222

11.2 Power analysis for simple hypothesis tests 224

11.3 Power analysis for linear regression models 228

11.4 Power analysis for log-likelihood regression models 229

11.5 Power analysis for hierarchical regression models 231

11.6 Power analysis using Python 232

12 Further Exercises for Practice   **235**

12.1 Analyzing graduate salaries 235

    12.1.1 The graduates data set 236

    12.1.2 Discussion questions 236

    12.1.3 Data exercises 236

12.2 Analyzing a recruiting process 237

    12.2.1 The recruiting data set 238

    12.2.2 Discussion questions 238

    12.2.3 Data exercises 239

12.3 Analyzing the drivers of performance ratings 239

    12.3.1 The employee_performance data set 240

    12.3.2 Discussion questions 240

    12.3.3 Data exercises 241

12.4 Analyzing promotion differences between groups 241

    12.4.1 The promotion data set 242

    12.4.2 Discussion questions 242

    12.4.3 Data exercises 242

12.5 Analyzing feedback on learning programs 243

    12.5.1 The learning data set 243

    12.5.2 Discussion questions 244

    12.5.3 Data exercises 244

References   **247**

Glossary   **249**

Index   **253**

Notes on data used in this book

For R and Python users, each of the data sets used in this book can be downloaded individually by following the code in each chapter. Alternatively for R users who intend to work through all of the chapters, all data sets can be loaded into an R session in advance by installing and loading the peopleanalyticsdata R package.

```r
# install peopleanalyticsdata package
install.packages("peopleanalyticsdata")
library(peopleanalyticsdata)

# see a list of data sets
data(package = "peopleanalyticsdata")

# find out more about a specific data set ('managers' example)
help(managers)
```

Foreword by Alexis Fink

Over the past decade or so, increases in compute power, emergence of friendly analytic tools and an explosion of data have created a wonderful opportunity to bring more analytical rigor to nearly every imaginable question. Not coincidentally, organizations are increasingly looking to apply all that data and capability to what is typically their greatest area of expense and their greatest strategic differentiator—their people. For too long, many of the most critical decisions in an organization—people decisions—had been guided by gut instinct or borrowed 'best practices' and the democratization of people analytics opened up enticing pathways to fix that. Suddenly, analysts who were originally interested in data problems began to be interested in people problems, and HR professionals who had dedicated their careers to solving people problems needed more sophisticated analysis and data storytelling to make their cases and to refine their approaches for greater efficiency, effectiveness and impact.

Doing data work with people in organizations has complexities that some other types of data work doesn't. Often, the employee populations are relatively smaller than data sets used in other areas, sometimes limiting the methods that can be used. Various regulatory requirements may dictate what data can be gathered and used, and what types of evidence might be required for various programs or people strategies. Human behavior and organizations are sufficiently complex that typically, multiple factors work together in influencing an outcome. Effects can be subtle or meaningful only in combination, or difficult to tease apart. While in many disciplines, prediction is the most important aim, for most people analytics projects and practitioners, understanding *why* something is happening is critical.

While the universe of analytical approaches is wonderful and vast, the best 'Swiss army knife' we have in people analytics is regression. This volume is an accessible, targeted work aimed directly at supporting professionals doing people analytics work. I've had the privilege of knowing and respecting Keith McNulty for many years – he is the rare and marvelous individual who is deeply expert in the mechanics of data and analytics, curious about and steeped in the opportunities to improve the effectiveness and well-being of people at work, and a gifted teacher and storyteller. He is among the most prolific standard-bearers for people analytics. This new open-source volume is in keeping with many years of contributions to the practice of understanding people at work.

DOI: 10.1201/9781003194156-0

After nearly 30 years of doing people analytics work and the privilege of leading people analytics teams at several leading global organizations, I am still excited by the problems we get to solve, the insights we get to spawn, and the tremendous impact we can have on organizations and the people that comprise them. This work is human and technical and important and exciting and deeply gratifying. I hope that you will find this *Handbook of Regression Modeling in People Analytics* helps you uncover new truths and create positive impacts in your own work.

Alexis A. Fink
December 2020

Alexis A. Fink, PhD *is a leading figure in people analytics and has led major people analytics teams at Microsoft and Intel before her current role as Vice President of People Analytics and Workforce Strategy at Facebook. She is a Fellow of the Society for Industrial and Organizational Psychology and is a frequent author, journal editor and research leader in her field.*

Introduction

As a fresh-faced undergraduate in mathematics in the 1990s, I took an introductory course in statistics in my first term. I would never take another. I struggled with the subject, scored my lowest grade in it and swore I would never go anywhere near it again.

How wrong I was. Today I live and breathe statistics. How did that happen?

Firstly, statistics is about solving real-world problems, and amazingly there was not a single mention of a relatable problem from real life in that course I took all those years ago, just abstract mathematics. Nowadays, I know from my work and my personal learning activities that the mathematics has no meaning without a motivating problem to apply it to, and you'll see example problems all through this book.

Secondly, statistics is all about data, and working with real data has encouraged me to reengage with statistics and come at it from a different angle—bottom-up you could say. Suddenly all those concepts that were put up on whiteboards using abstract formulas now had real meaning and consequence to the data I was working with. For me, real data helps statistical theory come to life, and this book is supported by numerous data sets designed for the reader to engage with.

But one more step solidified my newfound love of statistics, and that was when I put regression modeling into practice. Faced with data sets that I initially believed were just far too messy and random to be able to produce genuine insights, I progressively became more and more fascinated by how regression can cut through the messiness, compartmentalize the randomness and lead you straight to inferences that are often surprising both in their clarity and in their conclusions.

Hence my motivation for writing this book, which is to give others—whether working in people analytics or otherwise—a starting point for a practical learning of regression methods, with the hope that they will see immediate applications to their work and take advantage of a much-underused toolkit that provides strong support for evidence-based practice.

I am a mathematician who is now a practitioner of analytics. For this reason you should see that this book is neither afraid of nor obsessed with the mathematics of the methodologies covered. It is my general observation that

DOI: 10.1201/9781003194156-0

many students and practitioners make the mistake of trying to run multivariate models without even a basic understanding of the underlying mathematics of those models, and I find it very difficult to see how they can be credible in responding to a wide range of questions or critique about their work without such an understanding. That said, it is also not necessary for students and practitioners to understand the deepest levels of theory in order to be fluent in running and interpreting multivariate models. In this book I have tried to limit the mathematical exposition to a level that allows confident and fluent execution and interpretation.

I subscribe strongly to the principles of open source sharing of knowledge. If you want to reference the material in this book or use the exercises or data sets in trainings or classes, you are free to do so and you do not need to request my permission. I only ask that you make reference to this book as the source.

I expect this book to improve over time. If you found this book or any part of it helpful to solving a problem, I'd love to hear about it. If you have comments to improve or question any aspect of the contents of this book I encourage you to leave an issue[1] on its Github repository. This is the most reliable way for me to see your comment. I promise to consider all comments and input, but I do have to make a personal judgment about whether they are helpful to the aims and purpose of this book. If I do make changes or additions based on your input I will make a point to acknowledge your contribution in future editions.

I would like to thank the following individuals who have reviewed or contributed to this book at some point during its development: Liz Romero, Alex LoPilato, Kevin Jaggs, Seth Saavedra. My sincere thanks to Alexis Fink for drawing on her years of people analytics experience to set the context for this book in her foreword. My thanks to the people analytics community for their constant encouragement and support in sharing theory, content and method, and to the R community for all the work they do in giving us amazing and constantly improving statistical tools to work with. Finally, I would like to thank my family for their patience and understanding on the evenings and weekends I dedicated to the writing of this book, and for tolerating far too much dinner conversation on the topic of statistics.

Keith McNulty
December 2020

---

[1] https://github.com/keithmcnulty/peopleanalytics-regression-book/issues

1

The Importance of Regression in People Analytics

In the 19th century, when Francis Galton first used the term 'regression' to describe a statistical phenomenon (see Chapter 4), little did he know how important that term would be today. Many of the most powerful tools of statistical inference that we now have at our disposal can be traced back to the types of early analysis that Galton and his contemporaries were engaged in. The sheer number of different regression-related methodologies and variants that are available to researchers and practitioners today is mind-boggling, and there are still rich veins of ongoing research that are focused on defining and refining new forms of regression to tackle new problems.

Neither could Galton have imagined the advent of the age of data we now live in. Those of us (like me) who entered the world of work even as recently as 20 years ago remember a time when most problems could not be expected to be solved using a data-driven approach, because there simply was no data. Things are very different now, with data being collected and processed all around us and available to use as direct or indirect measures of the phenomena we are interested in.

Along with the growth in data that we have seen in recent years, we have also seen a rapid growth in the availability of statistical tools—open source and free to use—that fundamentally change how we go about analytics. Gone are the clunky, complex, repeated steps on calculators or spreadsheets. In their place are lean statistical programming languages that can implement a regression analysis in milliseconds with a single line of code, allowing us to easily run and reproduce multivariate analysis at scale.

So given that we have access to well-developed methodology, rich sources of data and readily accessible tools, it is somewhat surprising that many analytics practitioners have a limited knowledge and understanding of regression and its applications. The aim of this book is to encourage inexperienced analytics practitioners to 'dip their toes' further into the wide and varied world of regression in order to deliver more targeted and precise insights to their organizations and stakeholders on the problems they are most interested in. While the primary subject matter focus of this book is the analysis of people-related phenomena, the material is easily and naturally transferable to other

DOI: 10.1201/9781003194156-1

disciplines. Therefore this book can be regarded as a practical introduction to a wide range of regression methods for any analytics student or practitioner.

It is my firm belief that all people analytics professionals should have a strong understanding of regression models and how to implement and interpret them in practice, and my aim with this book is to provide those who need it with help in getting there. In this chapter we will set the scene for the technical learning in the remainder of the book by outlining the relevance of regression models in people analytics practice. We also touch on some general inferential modeling theory to set a context for later chapters, and we provide a preview of the contents, structure and learning objectives of this book.

1.1 Why is regression modeling so important in people analytics?

People analytics involves the study of the behaviors and characteristics of people or groups in relation to important business, organizational or institutional outcomes. This can involve both qualitative methods and quantitative methods, but if data is available related to a particular topic of interest, then quantitative methods are almost always considered important. With such a specific focus on outcomes, any analyst working in people analytics will frequently need to model these outcomes both to understand what influences them and to potentially predict them in the future.

Modeling an outcome with the primary goal of understanding what influences it can be quite a different matter to modeling an outcome with the primary goal of predicting if it will happen in the future. If we need to understand what influences an outcome, we need to get inside a model and construct a formula or structure to infer how each variable acts on that outcome, we need to get a sense of which variables are meaningful or not, and we need to quantify the 'explainability' of the outcome based on our variables. If our primary aim is to predict the outcome, getting inside the model is less important because we don't have to explain the outcome, we just need to be confident that it predicts accurately.

A model constructed to understand an outcome is often called an *inferential* model. Regression models are the most well-known and well-used inferential models available, providing a wide range of measures and insights that help us explain the relationship between our input variables and our outcome of interest, as we shall see in later chapters of this book.

The current reality in the field of people analytics is that inferential models are more required than predictive models. There are two reasons for this.

First, data sets in people analytics are rarely large enough to facilitate satisfactory prediction accuracy, and so attention is usually shifted to inference for this reason alone. Second, in the field of people analytics, decisions often have a real impact on individuals. Therefore, even in the rare situations where accurate predictive modeling is attainable, stakeholders are unlikely to trust the output and bear the consequences of predictive models without some sort of elementary understanding of how the predictions are generated. This requires the analyst to consider inference power as well as predictive accuracy in selecting their modeling approach. Again, many regression models come to the fore because they are commonly able to provide both inferential and predictive value.

Finally, the growing importance of evidence-based practice in many clinical and professional fields has generated a need for more advanced modeling skills to satisfy rising demand for quantitative evidence from decision makers. In people-related fields such as human resources, many varieties of specialized regression-based models such as survival models or latent variable models have crossed from academic and clinical settings into business settings in recent years, and there is an increasing need for qualified individuals who understand and can implement and interpret these models in practice.

1.2 What do we mean by 'modeling'?

The term 'modeling' has a very wide range of meaning in everyday life and work. In this book we are focused on inferential modeling, and we define that as a specific form of statistical learning, which tries to discover and understand a mathematical relationship between a set of measurements of certain constructs and a measurement of an outcome of interest, based on a sample of data on each. Modeling is both a concept and a process.

1.2.1 The theory of inferential modeling

We will start with a theoretical description and then provide a real example from a later chapter to illustrate.

Imagine we have a population $\mathcal{P}$ for which we believe there may be a non-random relationship between a certain construct or set of constructs $\mathcal{C}$ and a certain measurable outcome $\mathcal{O}$. Imagine that for a certain sample S of observations from $\mathcal{P}$, we have a collection of data which we believe measure $\mathcal{C}$ to some acceptable level of accuracy, and for which we also have a measure of the outcome $\mathcal{O}$.

By convention, we denote the set of data that measure $\mathcal{C}$ on our sample S as $X = x_1, x_2, \ldots, x_p$, where each x_i is a vector (or column) of data measuring at least one of the constructs in $\mathcal{C}$. We denote the set of data that measure $\mathcal{O}$ on our sample set S as y. An upper-case X is used because the expectation is that there will be several columns of data measuring our constructs, and a lower-case y is used because the expectation is that the outcome is a single column.

Inferential modeling is the process of learning about a relationship (or lack of relationship) between the data in X and y and using that to *describe* a relationship (or lack of relationship) between our constructs $\mathcal{C}$ and our outcome $\mathcal{O}$ that is valid to a high degree of statistical certainty on the population $\mathcal{P}$. This process may include:

- Testing a proposed mathematical relationship in the form of a function, structure or iterative method
- Comparing that relationship against other proposed relationships
- Describing the relationship statistically
- Determining whether the relationship (or certain elements of it) can be generalized from the sample set S to the population $\mathcal{P}$

When we test a relationship between X and y, we acknowledge that data and measurements are imperfect and so each observation in our sample S may contain random error that we cannot control. Therefore we define our relationship as:

$$y = f(X) + \epsilon$$

where f is some transformation or function of the data in X and ϵ is a random, uncontrollable error.

f can take the form of a predetermined function with a formula defined on X, like a linear function for example. In this case we can call our model a *parametric model*. In a parametric model, the modeled value of y is known as soon as we know the values of X by simply applying the formula. In a non-parametric model, there is no predetermined formula that defines the modeled value of y purely in terms of X. Non-parametric models need further information in addition to X in order to determine the modeled value of y—for example the value of y in other observations with similar X values.

Regression models are designed to derive f using estimation based on statistical likelihood and expectation, founded on the theory of the distribution of random variables. Regression models can be both parametric and non-parametric, but by far the most commonly used methods (and the majority of those featured in this book) are parametric. Because of their foundation in statistical likelihood and expectation, they are particularly suited to helping answer questions of generalizability—that is, to what extent can the relationship being observed in the sample S be inferred for the population $\mathcal{P}$, which is usually the driving force in any form of inferential modeling.

Note that there is a difference between establishing a statistical relationship between $\mathcal{C}$ and $\mathcal{O}$ and establishing a *causal relationship* between the two. This can be a common trap that inexperienced statistical analysts fall into when communicating the conclusions of their modeling. Establishing that a relationship exists between a construct and an outcome is a far cry from being able to say that one *causes* the other. This is the common truism that 'correlation does not equal causation'.

To bring our theory to life, consider the walkthrough example in Chapter 4 of this book. In this example, we discuss how to establish a relationship between the academic results of students in the first three years of their education program and their results in the fourth year. In this case, our population $\mathcal{P}$ is all past, present and future students who take similar examinations, and our sample S is the students who completed their studies in the past three years. $X = x_1, x_2, x_3$ are each of the three scores from the first three years, and y is the score in the fourth year. We test f to be a linear relationship, and we establish that such a relationship can be generalized to the entire population $\mathcal{P}$ with a substantial level of statistical confidence[1].

Almost all our work in this book will refer to the variables X as *input variables* and the variable y as the *outcome variable*. There are many other common terms for these which you may find in other sources—for example X are often known as independent variables or covariates while y is often known as a dependent or response variable.

1.2.2 The process of inferential modeling

Inferential modeling—regression or otherwise—is a process of numerous steps. Typically the main steps are:

1. Defining the outcome of interest $\mathcal{O}$ and the input constructs $\mathcal{C}$ based on a broader evidence-based objective
2. Confirming that $\mathcal{O}$ has reliable measurement data
3. Determining which data can be used to measure $\mathcal{C}$
4. Determining a sample S and collecting, refining and cleaning data.
5. Performing exploratory data analysis (EDA) and proposing a set of models to test for f
6. Putting the data in an appropriate format for each model

---

[1]We also determine that x_1 (the first-year examination score) plays no significant role in f and that introducing some non-linearity into f further improves the statistical accuracy of the inferred relationship.

7. Running the models
8. Interpreting the outputs and performing model diagnostics
9. Selecting an optimal model or models
10. Articulating the inferences that can be generalized to apply to $\mathcal{P}$

This book is primarily focused on steps 7–10 of this process[2]. That is not to say that steps 1–6 are not important. Indeed these steps are critical and often loaded with analytic traps. Defining the problem, collecting reliable measures and cleaning and organizing data are still the source of much pain and angst for analysts, but these topics are for another day.

1.3   The structure, system and organization of this book

The purpose of this book is to put inexperienced practitioners firmly on a path to the confident and appropriate use of regression techniques in their day-to-day work. This requires enough of an understanding of the underlying theory so that judgments can be made about results, but also a practical set of steps to help practitioners apply the most common regression methods to a variety of typical modeling scenarios in a reliable and reproducible way.

In most chapters, time is spent on the underlying mathematics. Not to the degree of an academic theorist, but enough to ensure that the reader can associate some mathematical meaning to the outputs of models. While it may be tempting to skip the math, I strongly recommend against it if you intend to be a high performer in your field. The best analysts are those who can genuinely understand what the numbers are telling them.

The statistical programming language R is used for most of the practical demonstration in each chapter. Because R is open source and particularly well geared to inferential statistics, it is an excellent choice for those whose work involves a lot of inferential analysis. In later chapters, we show implementations of all of the available methodologies in Python, which is also a powerful open source tool for this sort of work.

Each chapter involves a walkthrough example to illustrate the specific method and to allow the reader to replicate the analysis for themselves. The exercises at the end of each chapter are designed so that the reader can try the same method on a different data set, or a different problem on the same data set, to test their learning and understanding. In the final chapter, a series of data sets and exercises are provided with limited instruction in order to give the reader an opportunity to test their overall knowledge in selecting and applying

---

[2]The book also addresses Steps 5 and 6 in some chapters.

regression methods to a variety of people analytics data sets and problems. All in all, sixteen different data sets are used as walkthrough or exercise examples, and all of these data sets are fictitious constructions unless otherwise indicated. Despite the fiction, they are deliberately designed to present the reader with something resembling how the data might look in practice, albeit cleaner and more organized.

The chapters of this book are arranged as follows:

- Chapter 2 covers the basics of the R programming language for those who want to attempt to jump straight in to the work in subsequent chapters but have very little R experience. Experienced R programmers can skip this chapter.
- Chapter 3 covers the essential statistical concepts needed to understand multivariate regression models. It also serves as a tutorial in univariate and bivariate statistics illustrated with real data. If you need help developing a decent understanding of descriptive statistics, random distribution and hypothesis testing, this is an important chapter to study.
- Chapter 4 covers linear regression and in the course of that introduces many other foundational concepts. The walkthrough example involves modeling academic results from prior results. The exercises involve modeling income levels based on various work and demographic factors.
- Chapter 5 covers binomial logistic regression. The walkthrough example involves modeling promotion likelihood based on performance metrics. The exercises involve modeling charitable donation likelihood based on prior donation behavior and demographics.
- Chapter 6 covers multinomial regression. The walkthrough example and exercise involves modeling the choice of three health insurance products by company employees based on demographic and position data.
- Chapter 7 covers ordinal regression. The walkthrough example involves modeling in-game disciplinary action against soccer players based on prior discipline and other factors. The exercises involve modeling manager performance based on varied data.
- Chapter 8 covers modeling options for data with explicit or latent hierarchy. The first part covers mixed modeling and uses a model of speed dating decisions as a walkthrough and example. The second part covers structural equation modeling and uses a survey for a political party as a walkthrough example. The exercises involve modeling latent variables in an employee engagement survey.
- Chapter 9 covers survival analysis, Cox proportional hazard regression and frailty models. The chapter uses employee attrition as a walkthrough example and exercise.
- Chapter 10 outlines alternative technical approaches to regression modeling in both R and Python. Models from previous chapters are used to illustrate these alternative approaches.
- Chapter 11 covers power analysis, focusing in particular on estimating the

required minimum sample sizes in establishing meaningful inferences for both simple statistical tests and multivariate models. Examples related to experimental studies are used to illustrate, such as concurrent validity studies of selection instruments. Example implementations in R and Python are outlined.

- Chapter 12 is a set of problems and data sets which will allow the reader to practice the skills they have learned in this book and apply them to a variety of people analytics domains such as recruiting, performance, promotion, compensation and learning. Sets of discussion questions and data exercises will guide the reader through each problem, but these are designed in a way that encourages the independent selection and application of the methods covered in this book. These data sets, problems and exercises would suit as homework material for classes in statistical modeling or people analytics.

2

The Basics of the R Programming Language

Most of the work in this book is implemented in the R statistical programming language which, along with Python, is one of the two languages that I use in my day-to-day statistical analysis. Sample implementations in Python are also provided at various points in the book. I have made efforts to keep the code as simple as possible, and I have tried to avoid the use of too many external packages. For the most part, readers should see (especially in the earlier chapters) that code blocks are short and simple, relying wherever possible on base R functionality. No doubt there are neater and more effective ways to code some of the material in this book using a wide array of R packages—and some of these are illustrated in Chapter 10—but my priority has been to keep the code simple, consistent and easily reproducible.

For those who wish to follow the method and theory without the implementations in this book, there is no need to read this chapter. However, the style of this book is to use implementation to illustrate theory and practice, and so tolerance of many code blocks will be necessary as you read onward.

For those who wish to simply replicate the models as quickly as possible, full code is provided throughout this book by means of interspersed code blocks. Assuming all the required external packages have been installed, these code blocks should all be transportable and immediately usable. For those who are extra-inquisitive and want to explore how I constructed graphics used for illustration (for which code is usually not displayed), the best place to go is the Github repository[1] for this book.

This chapter is for those who wish to learn the methods in this book but do not know how to use R. However, it is not intended to be a full tutorial on R. There are many more qualified individuals and existing resources that would better serve that purpose—in particular I recommend Wickham and Grolemund (2016). It is recommended that you consult these resources and become comfortable with the basics of R before proceeding into the later chapters of this book. However, acknowledging that many will want to dive in sooner rather than later, this chapter covers the absolute basics of R that will allow the uninitiated reader to proceed with at least some orientation.

---

[1]https://github.com/keithmcnulty/peopleanalytics-regression-book

DOI: 10.1201/9781003194156-2

2.1   What is R?

R is a programming language that was originally developed by and for statisticians, but in recent years its capabilities and the environments in which
it is used have expanded greatly, with extensive use nowadays in academia
and the public and private sectors. There are many advantages to using a
programming language like R. Here are some:

1. It is completely free and open source.
2. It is faster and more efficient with memory than popular graphical
   user interface analytics tools.
3. It facilitates easier replication of analysis from person to person
   compared with many alternatives.
4. It has a large and growing global community of active users.
5. It has a large and rapidly growing universe of packages, which are
   all free and which provide the ability to do an extremely wide range
   of general and highly specialized tasks, statistical and otherwise.

There is often heated debate about which tools are better for doing non-
trivial statistical analysis. I personally find that R provides the widest array
of resources for those interested in inferential modeling, while Python has
a more well-developed toolkit for predictive modeling and machine learning.
Since the primary focus of this book is inferential modeling, the in-depth
walkthroughs are coded in R.

2.2   How to start using R

Just like most programming languages, R itself is an interpreter which receives
input and returns output. It is not very easy to use without an IDE. An IDE is
an *Integrated Development Environment*, which is a convenient user interface
allowing an R programmer to do all their main tasks including writing and
running R code, saving files, viewing data and plots, integrating code into
documents and many other things. By far the most popular IDE for R is
RStudio. An example of what the RStudio IDE looks like can be seen in
Figure 2.1.

To start using R, follow these steps:

1. Download and install the latest version of R from https://www.r-
   project.org/. Ensure that the version suits your operating system.

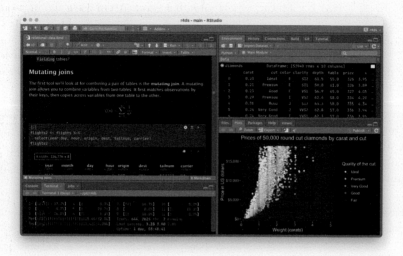

FIGURE 2.1: The RStudio IDE

2. Download the latest version of the RStudio IDE from
   https://rstudio.com/products/rstudio/ and view the video on
   that page to familiarize yourself with its features.

3. Open RStudio and play around.

The initial stages of using R can be challenging, mostly due to the need to
become familiar with how R understands, stores and processes data. Extensive
trial and error is a learning necessity. Perseverance is important in these early
stages, as well as an openness to seek help from others either in person or via
online forums.

2.3 Data in R

As you start to do tasks involving data in R, you will generally want to store
the things you create so that you can refer to them later. Simply calculating
something does not store it in R. For example, a simple calculation like this
can be performed easily:

```
3 + 3
```

```
## [1] 6
```

However, as soon as the calculation is complete, it is forgotten by R because the result hasn't been assigned anywhere. To store something in your R session, you will assign it a name using the <- operator. So I can assign my previous calculation to an object called my_sum, and this allows me to access the value at any time.

```
# store the result
my_sum <- 3 + 3

# now I can work with it
my_sum + 3
```

```
## [1] 9
```

You will see above that you can comment your code by simply adding a # to the start of a line to ensure that the line is ignored by the interpreter.

Note that assignment to an object does not result in the value being displayed. To display the value, the name of the object must be typed, the print() command used or the command should be wrapped in parentheses.

```
# show me the value of my_sum
my_sum
```

```
## [1] 6
```

```
# assign my_sum + 3 to new_sum and show its value
(new_sum <- my_sum + 3)
```

```
## [1] 9
```

2.3.1 Data types

All data in R has an associated type, to reflect the wide range of data that R is able to work with. The typeof() function can be used to see the type of a single scalar value. Let's look at the most common scalar data types.

Numeric data can be in integer form or double (decimal) form.

```r
# integers can be signified by adding an 'L' to the end
my_integer <- 1L
my_double <- 6.38

typeof(my_integer)
```

```
## [1] "integer"
```

```r
typeof(my_double)
```

```
## [1] "double"
```

Character data is text data surrounded by single or double quotes.

```r
my_character <- "THIS IS TEXT"
typeof(my_character)
```

```
## [1] "character"
```

Logical data takes the form TRUE or FALSE.

```r
my_logical <- TRUE
typeof(my_logical)
```

```
## [1] "logical"
```

2.3.2 Homogeneous data structures

Vectors are one-dimensional structures containing data of the same type and
are notated by using `c()`. The type of the vector can also be viewed using
the `typeof()` function, but the `str()` function can be used to display both the
contents of the vector and its type.

```r
my_double_vector <- c(2.3, 6.8, 4.5, 65, 6)
str(my_double_vector)
```

```
##  num [1:5] 2.3 6.8 4.5 65 6
```

Categorical data—which takes only a finite number of possible values—can
be stored as a factor vector to make it easier to perform grouping and manipulation.

```r
categories <- factor(
  c("A", "B", "C", "A", "C")
)

str(categories)
```

```
##  Factor w/ 3 levels "A","B","C": 1 2 3 1 3
```

If needed, the factors can be given order.

```r
# character vector
ranking <- c("Medium", "High", "Low")
str(ranking)
```

```
##  chr [1:3] "Medium" "High" "Low"
```

```r
# turn it into an ordered factor
ranking_factors <- ordered(
  ranking, levels = c("Low", "Medium", "High")
)

str(ranking_factors)
```

```
##  Ord.factor w/ 3 levels "Low"<"Medium"<..: 2 3 1
```

The number of elements in a vector can be seen using the length() function.

```
length(categories)
```

```
## [1] 5
```

Simple numeric sequence vectors can be created using shorthand notation.

```
(my_sequence <- 1:10)
```

```
##  [1]  1  2  3  4  5  6  7  8  9 10
```

If you try to mix data types inside a vector, it will usually result in *type coercion*, where one or more of the types are forced into a different type to ensure homogeneity. Often this means the vector will become a character vector.

```
# numeric sequence vector
vec <- 1:5
str(vec)
```

```
##  int [1:5] 1 2 3 4 5
```

```
# create a new vector containing vec and the character "hello"
new_vec <- c(vec, "hello")

# numeric values have been coerced into their character equivalents
str(new_vec)
```

```
##  chr [1:6] "1" "2" "3" "4" "5" "hello"
```

But sometimes logical or factor types will be coerced to numeric.

```
# attempt a mixed logical and numeric
mix <- c(TRUE, 6)

# logical has been converted to binary numeric (TRUE = 1)
str(mix)
```

```
##   num [1:2] 1 6
```

```
# try to add a numeric to our previous categories factor vector
new_categories <- c(categories, 1)

# categories have been coerced to background integer representations
str(new_categories)
```

```
##   num [1:6] 1 2 3 1 3 1
```

Matrices are two-dimensional data structures of the same type and are built from a vector by defining the number of rows and columns. Data is read into the matrix down the columns, starting left and moving right. Matrices are rarely used for non-numeric data types.

```
# create a 2x2 matrix with the first four integers
(m <- matrix(c(1, 2, 3, 4), nrow = 2, ncol = 2))
```

```
##        [,1] [,2]
## [1,]    1    3
## [2,]    2    4
```

Arrays are n-dimensional data structures with the same data type and are not used extensively by most R users.

2.3.3   Heterogeneous data structures

Lists are one-dimensional data structures that can take data of any type.

```
my_list <- list(6, TRUE, "hello")
str(my_list)
```

```
## List of 3
##  $ : num 6
##  $ : logi TRUE
##  $ : chr "hello"
```

List elements can be any data type and any dimension. Each element can be given a name.

```
new_list <- list(
  scalar = 6,
  vector = c("Hello", "Goodbye"),
  matrix = matrix(1:4, nrow = 2, ncol = 2)
)

str(new_list)
```

```
## List of 3
##  $ scalar: num 6
##  $ vector: chr [1:2] "Hello" "Goodbye"
##  $ matrix: int [1:2, 1:2] 1 2 3 4
```

Named list elements can be accessed by using $.

```
new_list$matrix
```

```
##      [,1] [,2]
## [1,]    1    3
## [2,]    2    4
```

Dataframes are the most used data structure in R; they are effectively a named list of vectors of the same length, with each vector as a column. As such, a dataframe is very similar in nature to a typical database table or spreadsheet.

```r
# two vectors of different types but same length
names <- c("John", "Ayesha")
ages <- c(31, 24)

# create a dataframe
(df <- data.frame(names, ages))
```

```
##     names ages
## 1    John   31
## 2 Ayesha   24
```

```r
# get types of columns
str(df)
```

```
## 'data.frame':    2 obs. of  2 variables:
##  $ names: chr   "John" "Ayesha"
##  $ ages : num  31 24
```

```r
# get dimensions of df
dim(df)
```

```
## [1] 2 2
```

2.4 Working with dataframes

The dataframe is the most common data structure used by analysts in R, due to its similarity to data tables found in databases and spreadsheets. We will work almost entirely with dataframes in this book, so let's get to know them.

2.4.1 Loading and tidying data in dataframes

To work with data in R, you usually need to pull it in from an outside source into a dataframe[2]. R facilitates numerous ways of importing data from simple

---

[2]R also has some built-in data sets for testing and playing with. For example, check out `mtcars` by typing it into the terminal, or type `data()` to see a full list of built-in data sets.

.csv files, from Excel files, from online sources or from databases. Let's load a data set that we will use later—the salespeople data set, which contains some information on the sales, average customer ratings and performance ratings of salespeople. The read.csv() function can accept a URL address of the file if it is online.

```
# url of data set
url <- "http://peopleanalytics-regression-book.org/data/salespeople.csv"

# load the data set and store it as a dataframe called salespeople
salespeople <- read.csv(url)
```

We might not want to display this entire data set before knowing how big it is. We can view the dimensions, and if it is too big to display, we can use the head() function to display just the first few rows.

```
dim(salespeople)
```

```
## [1] 351    4
```

```
# hundreds of rows, so view first few
head(salespeople)
```

```
##    promoted sales customer_rate performance
## 1         0   594          3.94           2
## 2         0   446          4.06           3
## 3         1   674          3.83           4
## 4         0   525          3.62           2
## 5         1   657          4.40           3
## 6         1   918          4.54           2
```

We can view a specific column by using $, and we can use square brackets to view a specific entry. For example if we wanted to see the 6th entry of the sales column:

```
salespeople$sales[6]
```

```
## [1] 918
```

Alternatively, we can use a [row, column] index to get a specific entry in the dataframe.

```
salespeople[34, 4]
```

```
## [1] 3
```

We can take a look at the data types using str().

```
str(salespeople)
```

```
## 'data.frame':    351 obs. of  4 variables:
##  $ promoted     : int  0 0 1 0 1 1 0 0 0 0 ...
##  $ sales        : int  594 446 674 525 657 918 318 364 342 387 ...
##  $ customer_rate: num  3.94 4.06 3.83 3.62 4.4 4.54 3.09 4.89 3.74 3 ...
##  $ performance  : int  2 3 4 2 3 2 3 1 3 3 ...
```

We can also see a statistical summary of each column using summary(), which tells us various statistics depending on the type of the column.

```
summary(salespeople)
```

```
##    promoted           sales        customer_rate    performance
##  Min.   :0.0000   Min.   :151.0   Min.   :1.000   Min.   :1.0
##  1st Qu.:0.0000   1st Qu.:389.2   1st Qu.:3.000   1st Qu.:2.0
##  Median :0.0000   Median :475.0   Median :3.620   Median :3.0
##  Mean   :0.3219   Mean   :527.0   Mean   :3.608   Mean   :2.5
##  3rd Qu.:1.0000   3rd Qu.:667.2   3rd Qu.:4.290   3rd Qu.:3.0
##  Max.   :1.0000   Max.   :945.0   Max.   :5.000   Max.   :4.0
##                   NA's   :1       NA's   :1       NA's   :1
```

Note that there is missing data in this dataframe, indicated by NAs in the summary. Missing data is identified by a special NA value in R. This should not be confused with "NA", which is simply a character string. The function is.na() will look at all values in a vector or dataframe and return TRUE or FALSE based on whether they are NA or not. By adding these up using the sum() function, it will take TRUE as 1 and FALSE as 0, which effectively provides a count of missing data.

```
sum(is.na(salespeople))
```

```
## [1] 3
```

This is a small number of NAs given the dimensions of our data set and we might want to remove the rows of data that contain NAs. The easiest way is to use the `complete.cases()` function, which identifies the rows that have no NAs, and then we can select those rows from the dataframe based on that condition. Note that you can overwrite objects with the same name in R.

```
salespeople <- salespeople[complete.cases(salespeople), ]

# confirm no NAs
sum(is.na(salespeople))
```

```
## [1] 0
```

We can see the unique values of a vector or column using the `unique()` function.

```
unique(salespeople$performance)
```

```
## [1] 2 3 4 1
```

If we need to change the type of a column in a dataframe, we can use the `as.numeric()`, `as.character()`, `as.logical()` or `as.factor()` functions. For example, given that there are only four unique values for the performance column, we may want to convert it to a factor.

```
salespeople$performance <- as.factor(salespeople$performance)
str(salespeople)
```

```
## 'data.frame':    350 obs. of  4 variables:
## $ promoted     : int  0 0 1 0 1 1 0 0 0 0 ...
## $ sales        : int  594 446 674 525 657 918 318 364 342 387 ...
## $ customer_rate: num  3.94 4.06 3.83 3.62 4.4 4.54 3.09 4.89 3.74 3 ...
## $ performance  : Factor w/ 4 levels "1","2","3","4": 2 3 4 2 3 2 3 1 3 3 ...
```

2.4.2   Manipulating dataframes

Dataframes can be subsetted to contain only rows that satisfy specific conditions.

```
(sales_720 <- subset(salespeople, subset = sales == 720))
```

```
##      promoted sales customer_rate performance
## 290        1   720          3.76           3
```

Note the use of ==, which is used in many programming languages, to test for precise equality. Similarly we can select columns based on inequalities (> for 'greater than', < for 'less than', >= for 'greater than or equal to', <= for 'less than or equal to', or != for 'not equal to'). For example:

```
high_sales <- subset(salespeople, subset = sales >= 700)
head(high_sales)
```

```
##     promoted sales customer_rate performance
## 6         1   918          4.54           2
## 12        1   716          3.16           3
## 20        1   937          5.00           2
## 21        1   702          3.53           4
## 25        1   819          4.45           2
## 26        1   736          3.94           4
```

To select specific columns use the select argument.

```
salespeople_sales_perf <- subset(salespeople,
                        select = c("sales", "performance"))
head(salespeople_sales_perf)
```

```
##    sales performance
## 1    594           2
## 2    446           3
## 3    674           4
## 4    525           2
## 5    657           3
## 6    918           2
```

Two dataframes with the same column names can be combined by their rows.

```
low_sales <- subset(salespeople, subset = sales < 400)

# bind the rows of low_sales and high_sales together
low_and_high_sales = rbind(low_sales, high_sales)
head(low_and_high_sales)
```

```
##    promoted sales customer_rate performance
## 7         0   318          3.09           3
## 8         0   364          4.89           1
## 9         0   342          3.74           3
## 10        0   387          3.00           3
## 15        0   344          3.02           2
## 16        0   372          3.87           3
```

Two dataframes with different column names can be combined by their columns.

```
# two dataframes with two columns each
sales_perf <- subset(salespeople,
                 select = c("sales", "performance"))
prom_custrate <- subset(salespeople,
                 select = c("promoted", "customer_rate"))

# bind the columns to create a dataframe with four columns
full_df <- cbind(sales_perf, prom_custrate)
head(full_df)
```

```
##   sales performance promoted customer_rate
## 1   594           2        0          3.94
## 2   446           3        0          4.06
## 3   674           4        1          3.83
## 4   525           2        0          3.62
## 5   657           3        1          4.40
## 6   918           2        1          4.54
```

2.5　Functions, packages and libraries

In the code so far we have used a variety of functions. For example `head()`, `subset()`, `rbind()`. Functions are operations that take certain defined inputs and return an output. Functions exist to perform common useful operations.

2.5.1　Using functions

Functions usually take one or more arguments. Often there are a large number of arguments that a function can take, but many are optional and not required to be specified by the user. For example, the function `head()`, which displays the first rows of a dataframe[3], has only one required argument x: the name of the dataframe. A second argument is optional, n: the number of rows to display. If n is not entered, it is assumed to have the default value n = 6.

When running a function, you can either specify the arguments by name or you can enter them in order without their names. If you enter arguments without naming them, R expects the arguments to be entered in exactly the right order.

```
# see the head of salespeople, with the default of six rows
head(salespeople)
```

```
##    promoted sales customer_rate performance
## 1         0   594          3.94           2
## 2         0   446          4.06           3
## 3         1   674          3.83           4
## 4         0   525          3.62           2
## 5         1   657          4.40           3
## 6         1   918          4.54           2
```

```
# see fewer rows - arguments need to be in the right order if not named
head(salespeople, 3)
```

---

[3]It actually has a broader definition but is mostly used for showing the first rows of a dataframe.

```
##    promoted sales customer_rate performance
## 1         0   594          3.94           2
## 2         0   446          4.06           3
## 3         1   674          3.83           4
```

```
# or if you don't know the right order,
# name your arguments and you can put them in any order
head(n = 3, x = salespeople)
```

```
##    promoted sales customer_rate performance
## 1         0   594          3.94           2
## 2         0   446          4.06           3
## 3         1   674          3.83           4
```

2.5.2 Help with functions

Most functions in R have excellent help documentation. To get help on the head() function, type help(head) or ?head. This will display the results in the Help browser window in RStudio. Alternatively you can open the Help browser window directly in RStudio and do a search there. An example of the browser results for head() is in Figure 2.2.

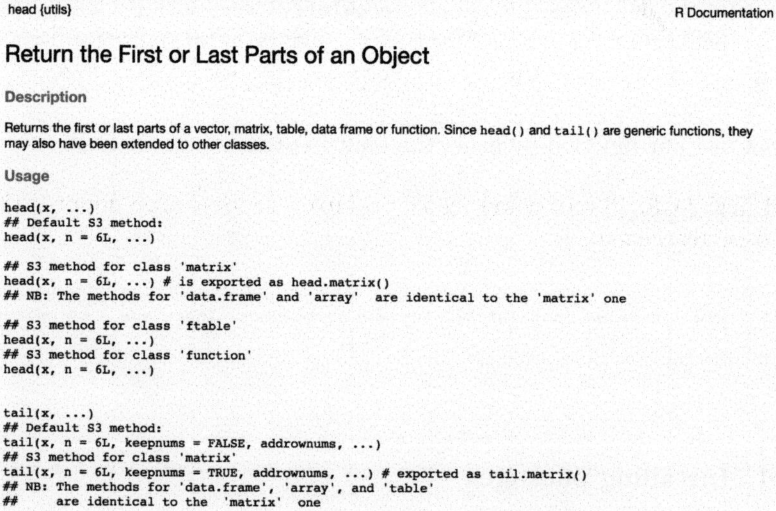

head (utils)   R Documentation

Return the First or Last Parts of an Object

Description

Returns the first or last parts of a vector, matrix, table, data frame or function. Since head() and tail() are generic functions, they may also have been extended to other classes.

Usage

```
head(x, ...)
## Default S3 method:
head(x, n = 6L, ...)

## S3 method for class 'matrix'
head(x, n = 6L, ...) # is exported as head.matrix()
## NB: The methods for 'data.frame' and 'array'  are identical to the 'matrix' one

## S3 method for class 'ftable'
head(x, n = 6L, ...)
## S3 method for class 'function'
head(x, n = 6L, ...)


tail(x, ...)
## Default S3 method:
tail(x, n = 6L, keepnums = FALSE, addrownums, ...)
## S3 method for class 'matrix'
tail(x, n = 6L, keepnums = TRUE, addrownums, ...) # exported as tail.matrix()
## NB: The methods for 'data.frame', 'array', and 'table'
##     are identical to the  'matrix' one
```

FIGURE 2.2: Results of a search for the head() function in the RStudio Help browser

The help page normally shows the following:

- Description of the purpose of the function
- Usage examples, so you can quickly see how it is used
- Arguments list so you can see the names and order of arguments
- Details or notes on further considerations on use
- Expected value of the output (for example `head()` is expected to return a similar object to its first input `x`)
- Examples to help orient you further (sometimes examples can be very abstract in nature and not so helpful to users)

2.5.3 Writing your own functions

Functions are not limited to those that come packaged in R. Users can write their own functions to perform tasks that are helpful to their objectives. Experienced programmers in most languages subscribe to a principle called DRY (Don't Repeat Yourself). Whenever a task needs to be done repeatedly, it is poor practice to write the same code numerous times. It makes more sense to write a function to do the task.

In this example, a simple function is written which generates a report on a dataframe:

```r
# create df_report function
df_report <- function(df) {
  paste("This dataframe contains", nrow(df), "rows and",
        ncol(df), "columns. There are", sum(is.na(df)), "NA entries.")
}
```

We can test our function by using the built-in `mtcars` data set in R.

```r
df_report(mtcars)
```

```
## [1] "This dataframe contains 32 rows and 11 columns. There are 0 NA entries."
```

2.5.4 Installing packages

All the common functions that we have used so far exist in the base R installation. However, the beauty of open source languages like R is that users can write their own functions or resources and release them to others via packages.

A package is an additional module that can be installed easily; it makes resources available which are not in the base R installation. In this book we will be using functions from both base R and from popular and useful packages. As an example, a popular package used for statistical modeling is the MASS package, which is based on methods in a popular applied statistics book[4].

Before an external package can be used, it must be installed into your package library using install.packages(). So to install MASS, type install.packages("MASS") into the console. This will send R to the main internet repository for R packages (known as CRAN). It will find the right version of MASS for your operating system and download and install it into your package library. If MASS needs other packages in order to work, it will also install these packages.

If you want to install more than one package, put the names of the packages inside a character vector—for example:

```
my_packages <- c("MASS", "DescTools", "dplyr")
install.packages(my_packages)
```

Once you have installed a package, you can see what functions are available by calling for help on it, for example using help(package = MASS). One package you may wish to install now is the peopleanalyticsdata package, which contains all the data sets used in this book. By installing and loading this package, all the data sets used in this book will be loaded into your R session and ready to work with. If you do this, you can ignore the read.csv() commands later in the book, which download the data from the internet.

2.5.5 Using packages

Once you have installed a package into your package library, to use it in your R session you need to load it using the library() function. For example, to load MASS after installing it, use library(MASS). Often nothing will happen when you use this command, but rest assured the package has been loaded and you can start to use the functions inside it. Sometimes when you load the package a series of messages will display, usually to make you aware of certain things that you need to keep in mind when using the package. Note that whenever you see the library() command in this book, it is assumed that you have already installed the package in that command. If you have not, the library() command will fail.

Once a package is loaded from your library, you can use any of the functions inside it. For example, the stepAIC() function is not available before you

---

[4]Venables and Ripley (2002)

load the MASS package but becomes available after it is loaded. In this sense, functions 'belong' to packages.

Problems can occur when you load packages that contain functions with the same name as functions that already exist in your R session. Often the messages you see when loading a package will alert you to this. When R is faced with a situation where a function exists in multiple packages you have loaded, R always defaults to the function in *the most recently loaded* package. This may not always be what you intended.

One way to completely avoid this issue is to get in the habit of *namespacing* your functions. To namespace, you simply use `package::function()`, so to safely call `stepAIC()` from MASS, you use `MASS::stepAIC()`. Most of the time in this book when a function is being called from a package outside base R, I use namespacing to call that function. This should help avoid confusion about which packages are being used for which functions.

2.5.6   The pipe operator

Even in the most elementary briefing about R, it is very difficult to ignore the pipe operator. The pipe operator makes code more natural to read and write and reduces the typical computing problem of many nested operations inside parentheses. The pipe operator comes inside many R packages, particularly magrittr and dplyr.

As an example, imagine we wanted to do the following two operations in one command:

1. Subset salespeople to only the sales values of those with sales less than 500
2. Take the mean of those values

In base R, one way to do this is:

```
mean(subset(salespeople$sales, subset = salespeople$sales < 500))
```

```
## [1] 388.6684
```

This is nested and needs to be read from the inside out in order to align with the instructions. The pipe operator %>% takes the command that comes before it and places it inside the function that follows it (by default as the first argument). This reduces complexity and allows you to follow the logic more clearly.

```
# load magrittr library to get the pipe operator
library(magrittr)

# use the pipe operator to lay out the steps more logically
subset(salespeople$sales, subset = salespeople$sales < 500) %>%
  mean()
```

```
## [1] 388.6684
```

This can be extended to perform arbitrarily many operations in one piped command.

```
salespeople$sales %>% # start with all data
  subset(subset = salespeople$sales < 500) %>% # get the subsetted data
  mean() %>% # take the mean value
  round() # round to the nearest integer
```

```
## [1] 389
```

The pipe operator is unique to R and is very widely used because it helps to make code more readable, it reduces complexity, and it helps orient around a common 'grammar' for the manipulation of data. The pipe operator helps you structure your code more clearly around nouns (objects), verbs (functions) and adverbs (arguments of functions). One of the most developed sets of packages in R that follows these principles is the `tidyverse` family of packages, which I encourage you to explore.

2.6 Errors, warnings and messages

As I mentioned earlier in this chapter, getting familiar with R can be frustrating at the beginning if you have never programmed before. You can expect to regularly see messages, warnings or errors in response to your commands. I encourage you to regard these as your friend rather than your enemy. It is very tempting to take the latter approach when you are starting out, but over time I hope you will appreciate some wisdom from my words.

Errors are serious problems which usually result in the halting of your code and a failure to return your requested output. They usually come with an indication of the source of the error, and these can sometimes be easy to understand and sometimes frustratingly vague and abstract. For example, an easy-to-understand error is:

```
subset(salespeople, subset = sales = 720)
```

```
Error: unexpected '=' in "subset(salespeople, subset = sales ="
```

This helps you see that you have used `sales = 720` as a condition to subset your data, when you should have used `sales == 720` for precise equality.

A much more challenging error to understand is:

```
head[salespeople]
```

```
Error in head[salespeople] : object of type 'closure' is not subsettable
```

When first faced with an error that you can't understand, try not to get frustrated and proceed in the knowledge that it usually can be fixed easily and quickly. Often the problem is much more obvious than you think, and if not, there is still a 99% likelihood that others have made this error and you can read about it online. The first step is to take a look at your code to see if you can spot what you did wrong. In this case, you may see that you have used square brackets `[]` instead of parentheses `()` when calling your `head()` function. If you cannot see what is wrong, the next step is to ask a colleague or do an internet search with the text of the error message you receive, or to consult online forums like `https://stackoverflow.com`. The more experienced you become, the easier it is to interpret error messages.

Warnings are less serious and usually alert you to something that you might be overlooking and which could indicate a problem with the output. In many cases you can ignore warnings, but sometimes they are an important reminder to go back and edit your code. For example, you may run a model which doesn't converge, and while this does not stop R from returning results, it is also very useful for you to know that it didn't converge.

Messages are pieces of information that may or may not be useful to you at a particular point in time. Sometimes you will receive messages when you load a package from your library. Sometimes messages will keep you up to date on the progress of a process that is taking a long time to execute.

2.7   Plotting and graphing

As you might expect in a well-developed programming language, there are numerous ways to plot and graph information in R. If you are doing exploratory data analysis on fairly simple data and you don't need to worry about pretty appearance or formatting, the built-in plot capabilities of base R are fine. If you need a pretty appearance, more precision, color coding or even 3D graphics or animation, there are also specialized plotting and graphing packages for these purposes. In general when working interactively in RStudio, graphical output will be rendered in the Plots pane, where you can copy it or save it as an image.

2.7.1   Plotting in base R

The simplest plot function in base R is plot(). This performs basic X-Y plotting. As an example, this code will generate a scatter plot of customer_rate against sales in the salespeople data set, with the results displayed in Figure 2.3. Note the use of the arguments main, xlab and ylab for customizing the axis labels and title for the plot.

```
# scatter plot of customer_rate against sales
plot(x = salespeople$sales, y = salespeople$customer_rate,
     xlab = "Sales ($m)", ylab = "Average customer rating",
     main = "Scatterplot of Sales vs Customer Rating")
```

Histograms of data can be generated using the hist() function. This command will generate a histogram of performance as displayed in Figure 2.4. Note the use of breaks to customize how the bars appear.

```
# convert performance ratings back to numeric data type for histogram
salespeople$performance <- as.numeric(salespeople$performance)

# histogram of performance ratings
hist(salespeople$performance, breaks = 0:4,
     xlab = "Performance Rating",
     main = "Histogram of Performance Ratings")
```

Box and whisker plots are excellent ways to see the distribution of a variable, and can be grouped against another variable to see bivariate patterns. For

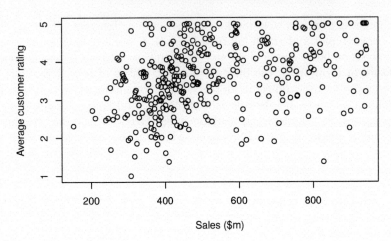

FIGURE 2.3: Simple scatterplot of `customer_rate` against `sales` in the `salespeople` data set

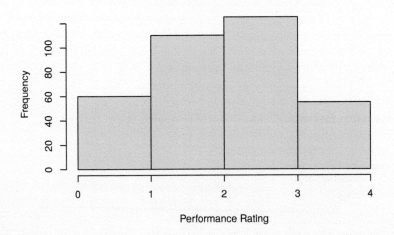

FIGURE 2.4: Simple histogram of `performance` in the `salespeople` data set

example, this command will show a box and whisker plot of `sales` grouped against `performance`, with the output shown in Figure 2.5. Note the use of the `formula` and `data` notation here to define the variable we are interested in and how we want it grouped. We will study this formula notation in greater depth later in this book.

```
# box plot of sales by performance rating
boxplot(formula = sales ~ performance, data = salespeople,
        xlab = "Performance Rating", ylab = "Sales ($m)",
        main = "Boxplot of Sales by Performance Rating")
```

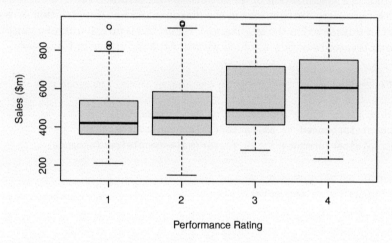

FIGURE 2.5: Simple box plot of `sales` grouped against `performance` in the `salespeople` data set

These are among the most common plots used for data exploration purposes. They are examples of a wider range of plotting and graphing functions available in base R, such as line plots, bar plots and other varieties which you may see later in this book.

2.7.2   Specialist plotting and graphing packages

By far the most commonly used specialist plotting and graphing package in R is `ggplot2`. `ggplot2` allows the flexible construction of a very wide range of charts and graphs, but uses a very specific command grammar which can take some getting used to. However, once learned, `ggplot2` can be an extremely powerful tool. Many of the illustratory figures used in this book are developed using `ggplot2` and while the code for these figures is generally not included for the sake of brevity, you can always find it in the source code of this book on Github[5]. A great learning resource for `ggplot2` is Wickham (2016).

The `plotly` package allows the use of the `plotly` graphing library in R. This is an excellent package for interactive graphing and is used for 3D illustrations in this book. Output can be rendered in HTML—allowing the user to play with and explore the graphs interactively—or can be saved as static 2D images.

`GGally` is a package that extends `ggplot2` to allow easy combination of charts and graphs. This is particularly valuable for quicker exploratory data analysis. One of its most popular functions is `ggpairs()`, which produces a pairplot. A pairplot is a visualization of all univariate and bivariate patterns in a data set, with univariate distributions in the diagonal and bivariate relationships or correlations displayed in the off-diagonal. Figure 2.6 is an example of a pairplot for the `salespeople` data set, which we will explore further in Chapter 5.

```r
library(GGally)

# convert performance and promotion to categorical
salespeople$promoted <- as.factor(salespeople$promoted)
salespeople$performance <- as.factor(salespeople$performance)

# pairplot of salespeople
GGally::ggpairs(salespeople)
```

2.8   Documenting your work using R Markdown

For anyone performing any sort of multivariate analysis using a statistical programming language, appropriate documentation and reproducibility of the work is essential to its success and longevity. If your code is not easily obtained or run by others, it is likely to have a very limited impact and lifetime. Learning how to create integrated documents that contain both text and code is critical to providing access to your code and narration of your work.

---

[5]`https://github.com/keithmcnulty/peopleanalytics-regression-book`

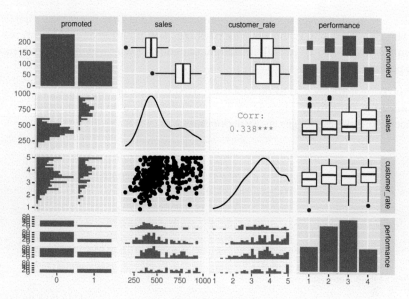

FIGURE 2.6: Pairplot of the `salespeople` data set

R Markdown is a package which allows you to create integrated documents containing both formatted text and executed code. It is, in my opinion, one of the best resources available currently for this purpose. This entire book has been created using R Markdown. You can start an R Markdown document in RStudio by installing the `rmarkdown` package and then opening a new R Markdown document file, which will have the suffix `.Rmd`.

R Markdown documents always start with a particular heading type called a YAML header, which contains overall information on the document you are creating. Care must be taken with the precise formatting of the YAML header, as it is sensitive to spacing and indentation. Usually a basic YAML header is created for you in RStudio when you start a new `.Rmd` file. Here is an example.

```
---
title: "My new document"
author: "Keith McNulty"
date: "25/01/2021"
output: html_document
---
```

The `output` part of this header has numerous options, but the most commonly used are `html_document`, which generates your document as a web page, and `pdf_document`, which generates your document as a PDF using the open source `LaTeX` software package. If you wish to create PDF documents you will need to have a version of LaTeX installed on your system. One R package that can do this for you easily is the tinytex package. The function install_tinytex() from this package will install a minimal version of LaTeX which is fine for most purposes.

R Markdown allows you to build a formatted document using many shorthand formatting commands. Here are a few examples of how to format headings and place web links or images in your document:

```
# My top heading

This section is about this general topic.

## My first sub heading

To see more information on this sub-topic visit [here](https://my.web.link).

## My second sub heading

Here is a nice picture about this sub-topic.

![](path/to/image)
```

Code can be written and executed and the results displayed inline using back-ticks. For example, writing

```
`r nrow(salespeople)`
```

inline will display 351 in the final document. Entire code blocks can be included and executed by using triple-backticks. The following code block:

```{r}
# show the first few rows of salespeople
head(salespeople)
```

will display this output:

```
##   promoted sales customer_rate performance
## 1        0   594          3.94           2
## 2        0   446          4.06           3
## 3        1   674          3.83           4
## 4        0   525          3.62           2
## 5        1   657          4.40           3
## 6        1   918          4.54           2
```

The {} wrapping allows you to specify different languages for your code chunk. For example, if you wanted to run Python code instead of R code you can use {python}. It also allows you to set options for the code chunk display separated by commas. For example, if you want the results of your code to be displayed, but without the code itself being displayed, you can use {r, echo = FALSE}.

The process of compiling your R Markdown code to produce a document is known as 'knitting.' To create a knitted document, you simply need to click on the 'Knit' button in RStudio that appears above your R Markdown code.

If you are not familiar with R Markdown, I strongly encourage you to learn it alongside R and to challenge yourself to write up any practice exercises you take on in this book using R Markdown. Useful cheat sheets and reference guides for R Markdown formatting and commands are available through the Cheatsheets section of the Help menu in RStudio. I also recommend Xie, Dervieux, and Riederer (2020) for a really thorough instruction and reference guide.

2.9 Learning exercises

2.9.1 Discussion questions

1. Describe the following data types: numeric, character, logical, factor.
2. Why is a vector known as a homogeneous data structure?
3. Give an example of a heterogeneous data structure in R.
4. What is the difference between NA and "NA"?
5. What operator is used to return named elements of a list and named columns of a dataframe?
6. Describe some functions that are used to manipulate dataframes.
7. What is a package and how do you install and use a new package?
8. Describe what is meant by 'namespacing' and why it might be useful.
9. What is the pipe operator, and why is it popular in R?
10. What is the difference between an error and a warning in R?
11. Name some simple plotting functions in base R.
12. Name some common specialist plotting and graphing packages in R.
13. What is R Markdown, and why is it useful to someone performing analysis using programming languages?

2.9.2　Data exercises

1. Create a character vector called `my_names` that contains all your first, middle and last names as elements. Calculate the length of `my_names`.

2. Create a second numeric vector called `which` which corresponds to `my_names`. The entries should be the position of each name in the order of your full name. Verify that it has the same length as `my_names`.

3. Create a dataframe called `names`, which consists of the two vectors `my_names` and `which` as columns. Calculate the dimensions of `names`.

4. Create a new dataframe `new_names` with the `which` column converted to character type. Verify that your command worked using `str()`.

5. Load the `ugtests` data set via the `peopleanalyticsdata` package or download it from the internet[6]. Calculate the dimensions of `ugtests` and view the first three rows only.

6. View a statistical summary of all of the columns of `ugtests`. Determine if there are any missing values.

7. View the subset of `ugtests` for values of `Yr1` greater than 50.

8. Install and load the package `dplyr`. Look up the help for the `filter()` function in this package and try to use it to repeat the task in the previous question.

9. Write code to find the mean of the `Yr1` test scores for all those who achieved `Yr3` test scores greater than 100. Round this mean to the nearest integer.

10. Familiarize yourself with the two functions `filter()` and `pull()` from `dplyr`. Use these functions to try to do the same calculation in the previous question using a single unbroken piped command. Be sure to namespace where necessary.

11. Create a scatter plot using the `ugtests` data with `Final` scores on the y axis and `Yr3` scores on the x axis.

12. Create your own 5-level grading logic and use it to create a new `finalgrade` column in the `ugtests` data set with grades 1–5 of increasing attainment based on the `Final` score in `ugtests`. Generate a histogram of this `finalgrade` column.

13. Using your new `ugtests` data with the extra column from the previous exercise, create a box plot of `Yr3` scores grouped by `finalgrade`.

14. Knit all of your answers to these exercises into an R Markdown document. Create one version that displays your code and answers, and another that just displays the answers.

---

[6]http://peopleanalytics-regression-book.org/data/ugtests.csv

3

Statistics Foundations

To properly understand multivariate models, an analyst needs to have a decent grasp of foundational statistics. Many of the assumptions and results of multivariate models require an understanding of these foundations in order to be properly interpreted. There are three topics that are particularly important for those proceeding further in this book:

1. Descriptive statistics of populations and samples
2. Distribution of random variables
3. Hypothesis testing

If you have never really studied these topics, I would strongly recommend taking a course in them and spending good time getting to know them. Again, just as the last chapter was not intended to be a comprehensive tutorial on R, neither is this chapter intended to be a comprehensive tutorial on introductory statistics. However, we will introduce some key concepts here that are critical to understanding later chapters, and as always we will illustrate using real data examples.

In preparation for this chapter we are going to download a data set that we will work through in a later chapter, and use it for practical examples and illustration purposes. The data are a set of information on the sales, customer ratings and performance ratings on a set of 351 salespeople as well as an indication of whether or not they were promoted.

```
# if needed, use online url to download salespeople data
url <- "http://peopleanalytics-regression-book.org/data/salespeople.csv"
salespeople <- read.csv(url)
```

Let's take a brief look at the first few rows of this data to make sure we know what is inside it.

```
head(salespeople)
```

DOI: 10.1201/9781003194156-3

```
##    promoted sales customer_rate performance
## 1         0   594          3.94           2
## 2         0   446          4.06           3
## 3         1   674          3.83           4
## 4         0   525          3.62           2
## 5         1   657          4.40           3
## 6         1   918          4.54           2
```

And let's understand the structure of this data.

```
str(salespeople)
```

```
## 'data.frame':    351 obs. of  4 variables:
##  $ promoted     : int  0 0 1 0 1 1 0 0 0 0 ...
##  $ sales        : int  594 446 674 525 657 918 318 364 342 387 ...
##  $ customer_rate: num  3.94 4.06 3.83 3.62 4.4 4.54 3.09 4.89 3.74 3 ...
##  $ performance  : int  2 3 4 2 3 2 3 1 3 3 ...
```

It looks like:

- promoted is a binary value, either 1 or 0, indicating 'promoted' or 'not promoted,' respectively.
- sales and customer_rate look like normal numerical values.
- performance looks like a set of performance categories—there appear to be four based on what we can see.

3.1 Elementary descriptive statistics of populations and samples

Any collection of numerical data on one or more variables can be described using a number of common statistical concepts. Let $x = x_1, x_2, ..., x_n$ be a sample of n observations of a variable drawn from a population.

3.1.1 Mean, variance and standard deviation

The **mean** is the average value of the observations and is defined by adding up all the values and dividing by the number of observations. The mean $\bar{x}$ of our sample x is defined as:

$$\bar{x} = \frac{1}{n} \sum_{i=1}^{n} x_i$$

While the mean of a sample x is denoted by $\bar{x}$, the mean of an entire population is usually denoted by μ. The mean can have a different interpretation depending on the type of data being studied. Let's look at the mean of three different columns of our salespeople data, making sure to ignore any missing data.

```
mean(salespeople$sales, na.rm = TRUE)
```

```
## [1] 527.0057
```

This looks very intuitive and appears to be the average amount of sales made by the individuals in the data set.

```
mean(salespeople$promoted, na.rm = TRUE)
```

```
## [1] 0.3219373
```

Given that this data can only have the value of 0 or 1, we interpret this mean as *likelihood* or *expectation* that an individual will be labeled as 1. That is, the average probability of promotion in the data set. If this data showed a perfectly random likelihood of promotion, we would expect this to take the value of 0.5. But it is lower than 0.5, which tells us that the majority of individuals are not promoted.

```
mean(salespeople$performance, na.rm = TRUE)
```

```
## [1] 2.5
```

Given that this data can only have the values 1, 2, 3 or 4, we interpret this as the *expected value* of the performance rating in the data set. Higher or lower means inform us about the distribution of the performance ratings. A low mean will indicate a skew towards a low rating, and a high mean will indicate a skew towards a high rating.

Other common statistical summary measures include the *median*, which is the middle value when the values are ranked in order, and the *mode*, which is the most frequently occurring value.

The **variance** is a measure of how much the data varies around its mean. There are two different definitions of variance. The **population variance** assumes that that we are working with the entire population and is defined as the average squared difference from the mean:

$$\text{Var}_p(x) = \frac{1}{n} \sum_{i=1}^{n} (x_i - \bar{x})^2$$

The **sample variance** assumes that we are working with a sample and attempts to estimate the variance of a larger population by applying *Bessel's correction* to account for potential sampling error. The sample variance is:

$$\text{Var}_s(x) = \frac{1}{n-1} \sum_{i=1}^{n} (x_i - \bar{x})^2$$

You can see that

$$\text{Var}_p(x) = \frac{n-1}{n} \text{Var}_s(x)$$

So as the data set gets larger, the sample variance and the population variance become less and less distinguishable, which intuitively makes sense.

Because we rarely work with full populations, the sample variance is calculated by default in R and in many other statistical software packages.

```r
# sample variance
(sample_variance_sales <- var(salespeople$sales, na.rm = TRUE))
```

```
## [1] 34308.11
```

So where necessary, we need to apply a transformation to get the population variance.

```r
# population variance (need length of non-NA data)
n <- length(na.omit(salespeople$sales))
(population_variance_sales <- ((n-1)/n) * sample_variance_sales)
```

```
## [1] 34210.09
```

Variance does not have intuitive scale relative to the data being studied, because we have used a 'squared distance metric', therefore we can square-root it to get a measure of 'deviance' on the same scale as the data. We call this the *standard deviation* $\sigma(x)$, where $\text{Var}(x) = \sigma(x)^2$. As with variance, standard deviation has both population and sample versions, and the sample version is calculated by default. Conversion between the two takes the form

$$\sigma_p(x) = \sqrt{\frac{n-1}{n}} \sigma_s(x)$$

```
# sample standard deviation
(sample_sd_sales <- sd(salespeople$sales, na.rm = TRUE))
```

```
## [1] 185.2245
```

```
# verify that sample sd is sqrt(sample var)
sample_sd_sales == sqrt(sample_variance_sales)
```

```
## [1] TRUE
```

```
# calculate population standard deviation
(population_sd_sales <- sqrt((n-1)/n) * sample_sd_sales)
```

```
## [1] 184.9597
```

Given the range of sales is [151, 945] and the mean is 527, we see that the standard deviation gives a more intuitive sense of the 'spread' of the data relative to its inherent scale.

3.1.2 Covariance and correlation

The **covariance** between two variables is a measure of the extent to which one changes as the other changes. If $y = y_1, y_2, \ldots, y_n$ is a second variable, and $\bar{x}$ and $\bar{y}$ are the means of x and y, respectively, then the **sample covariance** of x and y is defined as

$$\text{cov}_s(x,y) = \frac{1}{n-1}\sum_{i=1}^{n}(x_i - \bar{x})(y_i - \bar{y})$$

and as with variance, the **population covariance** is

$$\text{cov}_p(x,y) = \frac{n-1}{n}\text{cov}_s(x,y)$$

Again, the sample covariance is the default in R, and we need to transform to obtain the population covariance.

```
# get sample covariance for sales and customer_rate,
# ignoring observations with missing data
(sample_cov <- cov(salespeople$sales, salespeople$customer_rate,
                   use = "complete.obs"))
```

```
## [1] 55.81769
```

```
# convert to population covariance (need number of complete obs)
cols <- subset(salespeople, select = c("sales", "customer_rate"))
n <- nrow(cols[complete.cases(cols), ])
(population_cov <- ((n-1)/n) * sample_cov)
```

```
## [1] 55.65821
```

As can be seen, the difference in covariance is very small between the sample and population versions, and both confirm a positive relationship between sales and customer rating. However, we again see this issue that there is no intuitive sense of scale for this measure.

Pearson's correlation coefficient divides the covariance by the product of the standard deviations of the two variables:

$$r_{x,y} = \frac{\text{cov}(x,y)}{\sigma(x)\sigma(y)}$$

This creates a scale of -1 to 1 for $r_{x,y}$, which is an intuitive way of understanding both the direction and strength of the relationship between x and y, with -1 indicating that x increases perfectly as y decreases, 1 indicating that x increases perfectly as y increases, and 0 indicating that there is no relationship between the two.

As before, there is a sample and population version of the correlation coefficient, and R calculates the sample version by default. Similar transformations can be used to determine a population correlation coefficient and over large samples the two measures converge.

```r
# calculate sample correlation between sales and customer_rate
cor(salespeople$sales, salespeople$customer_rate, use = "complete.obs")
```

```
## [1] 0.337805
```

This tells us that there is a moderate positive correlation between sales and customer rating.

You will notice that we have so far used two variables on a continuous scale to demonstrate covariance and correlation. Pearson's correlation can also be used between a continuous scale and a dichotomous (binary) scale variable, and this is known as a **point-biserial correlation**.

```r
cor(salespeople$sales, salespeople$promoted, use = "complete.obs")
```

```
## [1] 0.8511283
```

Correlating ranked variables involves an adjusted approach leading to **Spearman's rho** (ρ) or **Kendall's tau** (τ), among others. We will not dive into the mathematics of this here, but a good source is Bhattacharya and Burman (2016). Spearman's or Kendall's variant should be used whenever at least one of the variables is a ranked variable, and both variants are available in R.

```r
# spearman's rho correlation
cor(salespeople$sales, salespeople$performance,
    method = "spearman", use = "complete.obs")
```

```
## [1] 0.2735446
```

```r
# kendall's tau correlation
cor(salespeople$sales, salespeople$performance,
    method = "kendall", use = "complete.obs")
```

[1] 0.2073609

In this case, both indicate a low to moderate correlation. Spearman's rho or Kendall's tau can also be used to correlate a ranked and a dichotomous variable, and this is known as a **rank-biserial correlation**.

3.2   Distribution of random variables

As we outlined in Section 1.2, when we build a model we are using a set of sample data to infer a general relationship on a larger population. A major underlying assumption in our inference is that we believe the real-life variables we are dealing with are random in nature. For example, we might be trying to model the drivers of the voting choice of millions of people in a national election, but we may only have sample data on a few thousand people. When we infer nationwide voting intentions from our sample, we assume that the characteristics of the voting population are random variables.

3.2.1   Sampling of random variables

When we describe variables as random, we are assuming that they take a form which is *independent and identically distributed*. Using our salespeople data as an example, we are assuming that the sales of one person in the data set is not influenced by the sales of another person in the data set. In this case, this seems like a reasonable assumption, and we will be making it for many (though not all) of the statistical methods used in this book. However, it is good to recognize that there are scenarios where this assumption cannot be made. For example, if the salespeople worked together in serving the same customers on the same products, and each individual's sales represented some proportion of the overall sales to the customer, we cannot say that the sales data is independent and identically distributed. In this case, we will expect to see some hierarchy in our data and will need to adjust our techniques accordingly to take this into consideration.

Under the central limit theorem, if we take samples from a random variable and calculate a summary statistic for each sample, that statistic is itself a random variable, and its mean converges to the true population statistic with more and more sampling. Let's test this with a little experiment on our salespeople data. Figure 3.1 shows the results of taking 10, 100 and 1000 different random samples of 50, 100 and 150 salespeople from the salespeople data set and creating a histogram of the resulting mean sales values. We can see how greater numbers of samples (down the rows) lead to a more normal distribution curve and larger sample sizes (across the columns) lead to a 'spikier' distribution with a smaller standard deviation.

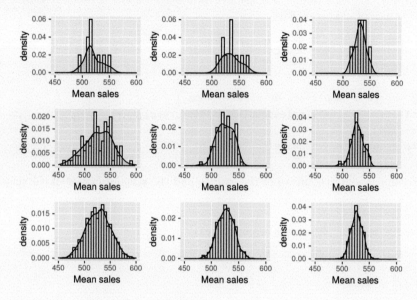

FIGURE 3.1: Histogram and density of mean sales from the salespeople data set based on sample sizes of 50, 100 and 150 (columns) and 10, 100 and 1000 samplings (rows)

3.2.2 Standard errors, the *t*-distribution and confidence intervals

One consequence of the observations in Figure 3.1 is that the summary statistics calculated from larger sample sizes fall into distributions that are 'narrower' and hence represent more precise estimations of the population statistic. The standard deviation of a sampled statistic is called the **standard error** of that statistic. In the special case of a sampled mean, the formula for the standard error of the mean can be derived to be

$$SE = \frac{\sigma}{\sqrt{n}}$$

where σ is the (sample) standard deviation and n is the sample size[1]. This confirms that the standard error of the mean decreases with greater sample

---

[1]Note that this formula assumes that the sample standard deviation is a close approximation of the population standard deviation, which is generally fine for samples that are not very small.

size, confirming our intuition that the estimation of the mean is more precise with larger samples.

To apply this logic to our `salespeople` data set, let's take a random sample of 100 values of `customer_rate`.

```
# set seed for reproducibility of sampling
set.seed(123)
```

```
# generate a sample of 100 observations
custrate <- na.omit(salespeople$customer_rate)
n <- 100
sample_custrate <- sample(custrate, n)
```

We can calculate the mean of the sample and the standard error of the mean.

```
# mean
(sample_mean <- mean(sample_custrate))
```

```
## [1] 3.6485
```

```
# standard error
(se <- sd(sample_custrate)/sqrt(n))
```

```
## [1] 0.08494328
```

Because the normal distribution is a frequency (or probability) distribution, we can interpret the standard error as a fundamental unit of 'sensitivity' around the sample mean. For greater multiples of standard errors around the sample mean, we can have greater certainty that the range contains the true population mean.

To calculate how many standard errors we would need around the sample mean to have a 95% probability of including the true population mean, we need to use the t-distribution. The t-distribution is essentially an approximation of the normal distribution acknowledging that we only have a sample estimate of the true population standard deviation in how we calculate the standard error. In this case where we are dealing with a single sample mean, we use the t-distribution with $n-1$ degrees of freedom. We can use the `qt()` function in R to find the standard error multiple associated with the level of certainty we need. In this case, we are looking for our true population mean to be outside the top 2.5% or bottom 2.5% of the distribution[2].

```
# get se multiple for 0.975
(t <- qt(p = 0.975, df = n - 1))
```

```
## [1] 1.984217
```

We see that approximately 1.98 standard errors on either side of our sample mean will give us 95% confidence that our range contains the true population mean. This is called the *95% confidence interval*[3].

```
# 95% confidence interval lower and upper bounds
lower_bound <- sample_mean - t*se
upper_bound <- sample_mean + t*se

cat(paste0('[', lower_bound, ', ', upper_bound, ']'))
```

```
## [3.47995410046738, 3.81704589953262]
```

3.3 Hypothesis testing

Observations about the distribution of statistics on samples of random variables allow us to construct tests for hypotheses of difference or similarity. Such hypothesis testing is useful in itself for simple bivariate analysis in practice settings, but it will be particularly critical in later chapters in determining whether models are useful or not. Before we go through some technical examples of hypothesis testing, let's overview the logic and intuition for how hypothesis testing works.

---

[2]As sample sizes increase and sample statistics get very close to population statistics, whether we use a *t*-distribution or a *z*-distribution (normal distribution) for determining confidence intervals or p-values becomes less important as they become almost identical on large samples. The output of some later models will refer to *t*-statistics and others to *z*-statistics, but the difference is only likely to matter in small samples of less than 50 or so observations. In this chapter we will use the *t*-distribution as it is a better choice for all sample sizes.

[3]Often we can use a rough estimate for larger samples that the 95% confidence interval is 2 standard errors either side of the sample mean.

The purpose of hypothesis testing is to establish a high degree of statistical certainty regarding a claim of difference in a population based on the properties of a sample. Consistent with a high burden of proof, we start from the hypothesis that there is no difference, called the *null hypothesis*. We only reject the null hypothesis if the statistical properties of the sample data render it very unlikely, in which case we confirm the *alternative hypothesis* that a statistical difference does exist in the population.

Most hypothesis tests can return a p-value, which is the maximum probability of finding the sample results (or results that are more extreme or unusual than the sample results) when the null hypothesis is true for the population. The analyst must decide on the level of p-value needed to reject the null hypothesis. This threshold is referred to as the significance level α (alpha). A common standard is to set α at 0.05. That is, we reject the null hypothesis if the p-value that we find for our sample results is less than 0.05. If we reject the null hypothesis at $\alpha = 0.05$, this means that the results we observe in the sample are so extreme or unusual that they would only occur by chance at most 1 in 20 times if the null hypothesis were true. An alpha of 0.05 is not the only standard of certainty used in research and practice, and in some fields of study smaller alphas are the norm, particularly if erroneous conclusions might have very serious consequences.

Three of the most common types of hypothesis tests are[4]:

1. Testing for a difference in the means of two groups
2. Testing for a non-zero correlation between two variables
3. Testing for a difference in frequency distributions between different categories

We will go through an example of each of these. In each case, you will see a three-step process. First, we calculate a test statistic. Second, we determine an expected distribution for that test statistic. Finally, we determine where our calculated statistic falls in that distribution in order to assess the likelihood of our sample occurring if the null hypothesis is true. During these examples, we will go through all the logic and calculation steps needed to do the hypothesis testing, before we demonstrate the simple functions that perform all the steps for you in R. Readers don't absolutely need to know all the details contained in this section, but a strong understanding of the underlying methods is encouraged.

---

[4] We go through these three examples both because they are relatively common and to illustrate the details of the logic behind hypothesis testing. By understanding how hypothesis tests work, this will allow the reader to grasp the meaning of other such tests like the F-test or the Wald test, which we will refer to in later chapters of this book

3.3.1 Testing for a difference in means (Welch's *t*-test)

Imagine that we are asked if, in general, the sales of low-performing salespeople are different from the sales of high-performing salespeople. This question refers to all salespeople, but we only have data for the sample in our `salespeople` data set. Let's take two subsets of our data for those with a performance rating of 1 and those with a performance rating of 4, and calculate the difference in mean sales.

```
# take two performance group samples
perf1 <- subset(salespeople, subset = performance == 1)
perf4 <- subset(salespeople, subset = performance == 4)

# calculate the difference in mean sales
(diff <- mean(perf4$sales) - mean(perf1$sales))
```

```
## [1] 154.9742
```

We can see that those with a higher performance rating in our sample did generate higher mean sales than those with a lower performance rating. But these are just samples, and we are being asked to give a conclusion about the populations they are drawn from. Let's take a null hypothesis that there is no difference in true mean sales between the two performance groups that these samples are drawn from. We combine the two samples and calculate the distribution around the difference in means. To *reject* the null hypothesis at $\alpha = 0.05$, we would need to determine that the 95% confidence interval of this distribution does not contain zero.

We calculate the standard error of the combined sample using the formula[5]:

$$\sqrt{\frac{\sigma_{\text{perf1}}^2}{n_{\text{perf1}}} + \frac{\sigma_{\text{perf4}}^2}{n_{\text{perf4}}}}$$

where σ_{perf1} and σ_{perf4} are the standard deviations of the two samples and n_{perf1} and n_{perf4} are the two sample sizes.

We use a special formula called the Welch-Satterthwaite approximation[6] to calculate the degrees of freedom for the two samples, which in this case calculates to 100.98[7]. This allows us to construct a 95% confidence interval for the difference between the means, and we can test whether this contains zero.

---

[5] If you are inquisitive about this formula, see the exercises at the end of this chapter.

[6] https://en.wikipedia.org/wiki/Welch-Satterthwaite_equation

[7] I've kept the gory details of how this is derived out of view, but you can see them if you view the source code for this book.

```
# calculate standard error of the two sets
se <- sqrt(sd(perf1$sales)^2/length(perf1$sales)
           + sd(perf4$sales)^2/length(perf4$sales))

# calculate the required t-statistic
t <- qt(p = 0.975, df = 100.98)

# calculate 95% confidence interval
(lower_bound <- diff - t*se)
```

```
## [1] 88.56763
```

```
(upper_bound <- diff + t*se)
```

```
## [1] 221.3809
```

```
# test if zero is inside this interval
(0 <= upper_bound) & (0 >= lower_bound)
```

```
## [1] FALSE
```

Since this has returned FALSE, we conclude that a mean difference of zero is outside the 95% confidence interval of our sample mean difference, and so we cannot have 95% certainty that the difference in population means is zero. We reject the null hypothesis that the mean sales of both performance levels are the same.

Looking at this graphically, we are assuming a *t*-distribution of the mean difference, and we are determining where zero sits in that distribution, as in Figure 3.2.

The red dashed lines in this diagram represent the 95% confidence interval around the mean difference of our two samples. The 'tails' of the curve outside of these two lines each represent a maximum of 0.025 probability for the true population mean. So we can see that the position of the blue dot-dashed line can correspond to a *maximum probability* that the population mean difference is zero. This is the *p-value* of the hypothesis test[8].

---

[8]We call this type of hypothesis test a *two-tailed* test, because the tested population mean can be either higher or lower than the sample mean, thus it can appear in any of the two tails for the null hypothesis to be rejected. *One-tailed* tests are used when you are testing for an alternative hypothesis that the difference is specifically 'less than zero' or 'greater than zero'. In the `t.test()` function in R, you can specify this in the arguments.

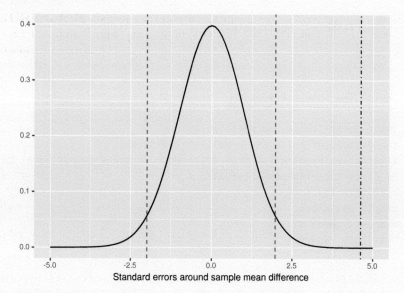

FIGURE 3.2: t-distribution of the mean sales difference between `perf1` and `perf4`, 95% confidence intervals (red dashed lines) and a zero difference (blue dot-dash line)

The p-value can be derived by calculating the standard error multiple associated with zero in the *t*-distribution (called the *t-statistic* or *t-value*), by applying the conversion function `pt()` to obtain the upper tail probability and then multiplying by 2 to get the probability associated with both tails of the distribution.

```r
# get t-statistic
t_actual <- diff/se
```

```r
# convert t-statistic to p-value
2*pt(t_actual, df = 100.98, lower = FALSE)
```

```
## [1] 1.093212e-05
```

Nowadays, it is never necessary to do these manual calculations ourselves because hypothesis tests are a standard part of statistical software. In R, the `t.test()` function performs a hypothesis test of difference in means of two samples and confirms our manually calculated p-value and 95% confidence interval.

```
t.test(perf4$sales, perf1$sales)
```

```
##
##  Welch Two Sample t-test
##
## data:  perf4$sales and perf1$sales
## t = 4.6295, df = 100.98, p-value = 1.093e-05
## alternative hypothesis: true difference in means is not equal to 0
## 95 percent confidence interval:
##    88.5676 221.3809
## sample estimates:
## mean of x mean of y
##   619.8909  464.9167
```

Because our p-value is less than our alpha of 0.05, we reject the null hypothesis in favor of the alternative hypothesis. The standard $\alpha = 0.05$ is associated with the term *statistically significant*. Therefore we could say here that the two performance groups have a statistically significant difference in mean sales.

In practice, there are numerous alphas that are of interest to analysts, each reflecting different levels of certainty. While 0.05 is the most common standard in many disciplines, more stringent alphas of 0.01 and 0.001 are often used in situations where a high degree of certainty is desirable (for example, some medical fields). Similarly, a less stringent alpha standard of 0.1 can be of interest particularly when sample sizes are small and the analyst is satisfied with 'indications' from the data. In many statistical software packages, including those that we will see in this book, tests that meet an $\alpha = 0.1$ standard are usually marked with period(.), those that meet $\alpha = 0.05$ with an asterisk($*$), $\alpha = 0.01$ a double asterisk($**$) and $\alpha = 0.001$ a triple asterisk($***$). Many leading statisticians have argued that p-values are more a test of sample size than anything else and have cautioned against too much of a focus on p-values in making statistical conclusions from data. In particular, situations where data and methodology have been deliberately manipulated to achieve certain alpha standards—a process known as 'p-hacking'—has been of increasing concern recently. See Chapter 11 for a better understanding of how the significance level and the sample size contribute to determining statistical power in hypothesis testing.

3.3.2 Testing for a non-zero correlation between two variables (t-test for correlation)

Imagine that we are given a sample of data for two variables and we are asked if the variables are correlated in the overall population. We can take a null

hypothesis that the variables are not correlated, determine a t-statistic associated with a zero correlation and convert this to a p-value. The t-statistic
associated with a correlation r between two samples of length n is often notated t^* and is defined as

$$t^* = \frac{r\sqrt{n-2}}{\sqrt{1-r^2}}$$

t^* can be converted to an associated p-value using a t-distribution in a similar
way to the previous section, this time with $n-2$ degrees of freedom in our
t-distribution. As an example, let's calculate t^* for the correlation between
sales and customer rating in our sample and convert it to a p-value.

```
# remove NAs from salespeople
salespeople <- salespeople[complete.cases(salespeople), ]

# calculate t_star
r <- cor(salespeople$sales, salespeople$customer_rate)
n <- nrow(salespeople)
t_star <- (r*sqrt(n - 2))/sqrt(1 - r^2)

# convert to p-value on t-distribution with n - 2 degrees of freedom
2*pt(t_star, df = n - 2, lower = FALSE)
```

```
## [1] 8.647952e-11
```

Again, there is a useful function in R to cut out the need for all our manual
calculations. The `cor.test()` function in R performs a hypothesis test on the
null hypothesis that two variables have zero correlation.

```
cor.test(salespeople$sales, salespeople$customer_rate)
```

```
##
##  Pearson's product-moment correlation
##
## data:  salespeople$sales and salespeople$customer_rate
## t = 6.6952, df = 348, p-value = 8.648e-11
## alternative hypothesis: true correlation is not equal to 0
## 95 percent confidence interval:
##   0.2415282 0.4274964
## sample estimates:
##       cor
## 0.337805
```

This confirms our manual calculations, and we see the null hypothesis has been rejected and we can conclude that there is a significant correlation between sales and customer rating.

3.3.3 Testing for a difference in frequency distribution between different categories in a data set (Chi-square test)

Imagine that we are asked if the performance category of each person in the salespeople data set has a relationship with their promotion likelihood. We will test the null hypothesis that there is no difference in the distribution of promoted versus not promoted across the four performance categories.

First we can produce a *contingency table*, which is a matrix containing counts of how many people were promoted or not promoted in each category.

```
# create contingency table of promoted vs performance
(contingency <- table(salespeople$promoted, salespeople$performance))
```

```
##
##      1  2  3  4
##   0 50 85 77 25
##   1 10 25 48 30
```

We can see by summing each row that for the total sample we can expect 113 people to be promoted and 237 to miss out on promotion. We can use this ratio to compute an expected proportion in each performance category under the assumption that the distribution was exactly the same across all four categories.

```
# calculate expected promoted and not promoted
(expected_promoted <- (sum(contingency[2, ])/sum(contingency)) *
    colSums(contingency))
```

```
##         1        2        3        4
## 19.37143 35.51429 40.35714 17.75714
```

```
(expected_notpromoted <- (sum(contingency[1, ])/sum(contingency)) *
    colSums(contingency))
```

```
##         1        2        3        4
## 40.62857 74.48571 84.64286 37.24286
```

Now we can compare our observed versus expected values using the difference metric:

$$\frac{(\text{observed} - \text{expected})^2}{\text{expected}}$$

and add these all up to get a total, known as the χ^2 statistic.

```
# calculate the difference metrics for promoted and not promoted
promoted <- sum((expected_promoted - contingency[2, ])^2/
                    expected_promoted)

notpromoted <- sum((expected_notpromoted - contingency[1, ])^2/
                    expected_notpromoted)

# calculate chi-squared statistic
(chi_sq_stat <- notpromoted + promoted)
```

```
## [1] 25.89541
```

The χ^2 statistic has an expected distribution that can be used to determine the p-value associated with this statistic. As with the t-distribution, the χ^2-distribution depends on the degrees of freedom. This is calculated by subtracting one from the number of rows and from the number of columns in the contingency table and multiplying them together. In this case we have 2 rows and 4 columns, which calculates to 3 degrees of freedom. Armed with our χ^2 statistic and our degrees of freedom, we can now calculate the p-value for the hypothesis test using the pchisq() function.

```
# calculate p-value from chi_squared stat
pchisq(chi_sq_stat, df = 3, lower.tail=FALSE)
```

```
## [1] 1.003063e-05
```

The chisq.test() function in R performs all the steps involved in a chi-square test of independence on a contingency table and returns the χ^2 statistic and associated p-value for the null hypothesis, in this case confirming our manual calculations.

```
chisq.test(contingency)
```

```
##
##  Pearson's Chi-squared test
##
## data:  contingency
## X-squared = 25.895, df = 3, p-value = 1.003e-05
```

Again, we can reject the null hypothesis and confirm the alternative hypothesis that there is a difference in the distribution of promoted/not promoted individuals between the four performance categories.

3.4 Foundational statistics in Python

Elementary descriptive statistics can be performed in Python using various packages. Descriptive statistics of numpy arrays are usually available as methods.

```python
import pandas as pd
import numpy as np

# get data
url = "http://peopleanalytics-regression-book.org/data/salespeople.csv"
salespeople = pd.read_csv(url)

# mean sales
mean_sales = salespeople.sales.mean()
print(mean_sales)
```

```
## 527.0057142857142
```

```python
# sample variance
var_sales = salespeople.sales.var()
print(var_sales)
```

```
## 34308.11458043389
```

```
# sample standard deviation
sd_sales = salespeople.sales.std()
print(sd_sales)
```

```
## 185.2244977869663
```

Population statistics can be obtained by setting the ddof parameter to zero.

```
# population standard deviation
popsd_sales = salespeople.sales.std(ddof = 0)
print(popsd_sales)
```

```
## 184.9597020864771
```

The numpy covariance function produces a covariance matrix.

```
# generate a sample covariance matrix between two variables
sales_rate = salespeople[['sales', 'customer_rate']]
sales_rate = sales_rate[~np.isnan(sales_rate)]
cov = sales_rate.cov()
print(cov)
```

```
##                     sales   customer_rate
## sales          34308.114580       55.817691
## customer_rate     55.817691        0.795820
```

Specific covariances between variable pairs can be pulled out of the matrix.

```
# pull out specific covariances
print(cov['sales']['customer_rate'])
```

```
## 55.817691199345006
```

Similarly for Pearson correlation:

```
# sample pearson correlation matrix
cor = sales_rate.corr()
print(cor)
```

```
##                    sales    customer_rate
## sales           1.000000        0.337805
## customer_rate   0.337805        1.000000
```

Specific types of correlation coefficients can be accessed via the stats module of the scipy package.

```
from scipy import stats
```

```
# spearman's correlation
stats.spearmanr(salespeople.sales, salespeople.performance,
nan_policy='omit')
```

```
## SpearmanrResult(correlation=0.27354459847452534, pvalue=2.00654343790
79837e-07)
```

```
# kendall's tau
stats.kendalltau(salespeople.sales, salespeople.performance,
nan_policy='omit')
```

```
## KendalltauResult(correlation=0.20736088105812, pvalue=2.73532582263
76615e-07)
```

Common hypothesis testing tools are available in scipy.stats. Here is an example of how to perform Welch's *t*-test on a difference in means of samples of unequal variance.

```
# get sales for top and bottom performers
perf1 = salespeople[salespeople.performance == 1].sales
perf4 = salespeople[salespeople.performance == 4].sales

# welch's t-test with unequal variance
ttest = stats.ttest_ind(perf4, perf1, equal_var=False)
print(ttest)
```

```
## Ttest_indResult(statistic=4.629477606844271, pvalue=1.0932443461577
038e-05)
```

As seen above, hypothesis tests for non-zero correlation coefficients are performed automatically as part of `scipy.stats` correlation calculations.

```
# calculate correlation and p-value
sales = salespeople.sales[~np.isnan(salespeople.sales)]

cust_rate = salespeople.customer_rate[
  ~np.isnan(salespeople.customer_rate)
]

cor = stats.pearsonr(sales, cust_rate)
print(cor)
```

```
## (0.33780504485867796, 8.647952212091035e-11)
```

Finally, a chi-square test of difference in frequency distribution can be performed on a contingency table as follows. The first value of the output is the χ^2 statistic, and the second value is the p-value.

```
# create contingency table for promoted versus performance
contingency = pd.crosstab(salespeople.promoted, salespeople.performance)

# perform chi-square test
chi2_test = stats.chi2_contingency(contingency)
print(chi2_test)
```

```
##              (25.895405268094862,              1.0030629464566802e-
05, 3, array([[40.62857143, 74.48571429, 84.64285714, 37.24285714],
##         [19.37142857, 35.51428571, 40.35714286, 17.75714286]]))
```

3.5   Learning exercises

3.5.1   Discussion questions

Where relevant in these discussion exercises, let $x = x_1, x_2, ..., x_n$ and $y = y_1, y_2, ..., y_m$ be samples of two random variables of length n and m respectively.

1. If the values of x can only take the form 0 or 1, and if their mean is 0.25, how many of the values equal 0?
2. If $m = n$ and $x + y$ is formed from the element-wise sum of x and y, show that the mean of $x + y$ is equal to the sum of the mean of x and the mean of y.
3. For a scalar multiplier a, show that $\text{Var}(ax) = a^2 \text{Var}(x)$.
4. Explain why the standard deviation of x is a more intuitive measure of the deviation in x than the variance.
5. Describe which two types of correlation you could use if x is an ordered ranking.
6. Describe the role of sample size and sampling frequency in the distribution of sampling means for a random variable.
7. Describe what a standard error of a statistic is and how it can be used to determine a confidence interval for the true population statistic.
8. If we conduct a t-test on the null hypothesis that x and y are drawn from populations with the same mean, describe what a p-value of 0.01 means.
9. **Extension:** The sum of variance law states that, for independent random variables x and y, $\text{Var}(x \pm y) = \text{Var}(x) + \text{Var}(y)$. Use this together with the identity from Exercise 3 to derive the formula for the standard error of the mean of $x = x_1, x_2, ..., x_n$:

$$SE = \frac{\sigma(x)}{\sqrt{n}}$$

10. **Extension:** In a similar way to Exercise 9, show that the standard error for the difference between the means of x and y is

$$\sqrt{\frac{\sigma(x)^2}{n} + \frac{\sigma(y)^2}{m}}$$

3.5.2 Data exercises

For these exercises, load the `charity_donation` data set via the `peopleanalyt-icsdata` package, or download it from the internet[9]. This data set contains information on a sample of individuals who made donations to a nature charity.

1. Calculate the mean `total_donations` from the data set.
2. Calculate the sample variance for `total_donation` and convert this to a population variance.
3. Calculate the sample standard deviation for `total_donations` and verify that it is the same as the square root of the sample variance.
4. Calculate the sample correlation between `total_donations` and `time_donating`. By using an appropriate hypothesis test, determine if these two variables are independent in the overall population.
5. Calculate the mean and the standard error of the mean for the first 20 entries of `total_donations`.
6. Calculate the mean and the standard error of the mean for the first 50 entries of `total_donations`. Verify that the standard error is less than in Exercise 5.
7. By using an appropriate hypothesis test, determine if the mean age of those who made a recent donation is different from those who did not.
8. By using an appropriate hypothesis test, determine if there is a difference in whether or not a recent donation was made according to where people reside.
9. **Extension:** By using an appropriate hypothesis test, determine if the age of those who have recently donated is at least 10 years older than those who have not recently donated in the population.
10. **Extension:** By using an appropriate hypothesis test, determine if the average donation amount is at least 10 dollars higher for those who recently donated versus those who did not. Retest for 20 dollars higher.

---

[9]http://peopleanalytics-regression-book.org/data/charity_donation.csv

4

Linear Regression for Continuous Outcomes

In this chapter, we will introduce and explore linear regression, one of the first learning methods to be developed by statisticians and one of the easiest to interpret. Despite its simplicity—indeed *because* of its simplicity—it can be a very powerful tool in many situations. Linear regression will often be the first methodology to be trialed on a given problem, and will give an immediate benchmark with which to judge the efficacy of other, more complex, modeling techniques. Given the ease of interpretation, many analysts will select a linear regression model over more complex approaches even if those approaches produce a slightly better fit. This chapter will also introduce many critical concepts that will apply to other modeling approaches as we proceed through this book. Therefore for inexperienced modelers this should be considered a foundational chapter which should not be skipped.

4.1 When to use it

4.1.1 Origins and intuition of linear regression

Linear regression, also known as *Ordinary Least Squares linear regression* or *OLS regression* for short, was developed independently by the mathematicians Gauss and Legendre at or around the first decade of the 19th century, and there remains today some controversy about who should take credit for its discovery. However, at the time of its discovery it was not actually known as 'regression'. This term became more popular following the work of Francis Galton—a British intellectual jack-of-all-trades and a cousin of Charles Darwin. In the late 1800s, Galton had researched the relationship between the heights of a population of almost 1000 children and the average height of their parents (mid-parent height). He was surprised to discover that there was not a perfect relationship between the height of a child and the average height of its parents, and that in general children's heights were more likely to be in a range that was closer to the mean for the total population. He described this statistical phenomenon as a 'regression towards mediocrity' ('regression' comes from a Latin term approximately meaning 'go back').

DOI: 10.1201/9781003194156-4

Figure 4.1 is a scatter plot of Galton's data with the black solid line showing what a perfect relationship would look like, the black dot-dashed line indicating the mean child height and the red dashed line showing the actual relationship determined by Galton[1]. You can regard the red dashed line as 'going back' from the perfect relationship (symbolized by the black line). This might give you an intuition that will help you understand later sections of this chapter. In an arbitrary data set, the red dashed line can lie anywhere between the black dot-dashed line (no relationship) and the black solid line (a perfect relationship). Linear regression is about finding the red dashed line in your data and using it to explain the degree to which your input data (the x axis) explains your outcome data (the y axis).

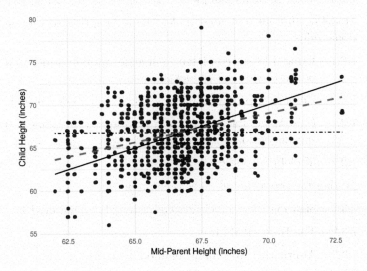

FIGURE 4.1: Galton's study of the height of children introduced the term 'regression'

4.1.2 Use cases for linear regression

Linear regression is particularly suited to a problem where the outcome of interest is on some sort of continuous scale (for example, quantity, money, height, weight). For outcomes of this type, it can be a first port of call before trying more complex modeling approaches. It is simple and easy to explain, and analysts will often accept a somewhat poorer fit using linear regression in order to avoid having to interpret a more complex model.

---

[1] This chart includes a random 'jitter' to better illustrate observations that are identical and was first used to illustrate Galton's data in Senn (2011).

Here are some illustrative examples of questions that could be tackled with a linear regression approach:

- Given a data set of demographic data, job data and current salary data, to what extent can current salary be explained by the rest of the data?

- Given annual test scores for a set of students over a four-year period, what is the relationship between the final test score and earlier test scores?

- Given a set of GPA data, SAT data and data on the percentile score on an aptitude test for a set of job applicants, to what extent can the GPA and SAT data explain the aptitude test score?

4.1.3 Walkthrough example

You are working as an analyst for the biology department of a large academic institution which offers a four-year undergraduate degree program. The academic leaders of the department are interested in understanding how student performance in the final-year examination of the degree program relates to performance in the prior three years.

To help with this, you have been provided with data for 975 individuals graduating in the past three years, and you have been asked to create a model to explain each individual's final examination score based on their examination scores for the first three years of their program. The Year 1 examination scores are awarded on a scale of 0–100, Years 2 and 3 on a scale of 0–200, and the Final year is awarded on a scale of 0–300.

We will load the ugtests data set into our session and take a brief look at it.

```
# if needed, download ugtests data
url <- "http://peopleanalytics-regression-book.org/data/ugtests.csv"
ugtests <- read.csv(url)
```

```
# look at the first few rows of data
head(ugtests)
```

```
##    Yr1 Yr2 Yr3 Final
## 1  27  50  52    93
## 2  70 104 126   207
## 3  27  36 148   175
## 4  26  75 115   125
## 5  46  77  75   114
## 6  86 122 119   159
```

The data looks as expected, with test scores for four years all read in as numeric data types, but of course this is only a few rows. We need a quick statistical and structural overview of the data.

```
# view structure
str(ugtests)
```

```
## 'data.frame':    975 obs. of  4 variables:
##  $ Yr1  : int  27 70 27 26 46 86 40 60 49 80 ...
##  $ Yr2  : int  50 104 36 75 77 122 100 92 98 127 ...
##  $ Yr3  : int  52 126 148 115 75 119 125 78 119 67 ...
##  $ Final: int  93 207 175 125 114 159 153 84 147 80 ...
```

```
# view statistical summary
summary(ugtests)
```

```
##       Yr1              Yr2              Yr3             Final
##  Min.   : 3.00   Min.   :  6.0   Min.   :  8.0   Min.   :  8
##  1st Qu.:42.00   1st Qu.: 73.0   1st Qu.: 81.0   1st Qu.:118
##  Median :53.00   Median : 94.0   Median :105.0   Median :147
##  Mean   :52.15   Mean   : 92.4   Mean   :105.1   Mean   :149
##  3rd Qu.:62.00   3rd Qu.:112.0   3rd Qu.:130.0   3rd Qu.:175
##  Max.   :99.00   Max.   :188.0   Max.   :198.0   Max.   :295
```

We can see that the results do seem to have different scales in the different years as we have been informed, and judging by the means, students seem to have found Year 2 exams more challenging. We can also be assured that there is no missing data, as these would have been displayed as NA counts in our summary if they existed.

We can also plot our four years of test scores pairwise to see any initial relationships of interest, as displayed in Figure 4.2.

```
library(GGally)

# display a pairplot of all four columns of data
GGally::ggpairs(ugtests)
```

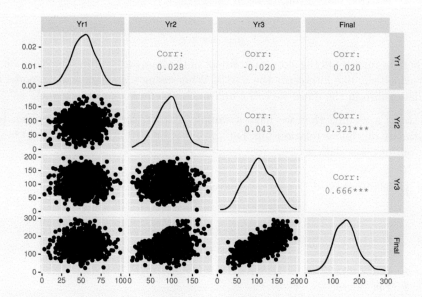

FIGURE 4.2: Pairplot of the `ugtests` data set

In the diagonal, we can see the distributions of the data in each column. We observe relatively normal-looking distributions in each year. We can see scatter plots and pairwise correlation statistics off the diagonal. For example, we see a particularly strong correlation between `Yr3` and `Final` test scores, a moderate correlation between `Yr2` and `Final` and relative independence elsewhere.

4.2   Simple linear regression

In order to visualize our approach and improve our intuition, we will start with *simple* linear regression, which is the case where there is only a single input variable and outcome variable.

4.2.1　Linear relationship between a single input and an outcome

Let our input variable be x and our outcome variable be y. Recalling the equation of a straight line, because we assume that the relationship is linear, we expect the relationship to be of the form:

$$y = mx + c$$

where m represents the slope or gradient of the line, and c represents the point at which the line intercepts the y axis. When using a straight line to model a relationship in the data, we call c and m the *coefficients* of the model.

Now let's assume that we have a sample of 10 observations with which to estimate our linear relationship. Let's take the first 10 values of Yr3 and Final in our ugtests data set:

```
(d <- head(ugtests[ , c("Yr3", "Final")], 10))
```

```
##      Yr3 Final
## 1     52    93
## 2    126   207
## 3    148   175
## 4    115   125
## 5     75   114
## 6    119   159
## 7    125   153
## 8     78    84
## 9    119   147
## 10    67    80
```

We can do a simple plot of these observations as in Figure 4.3. Intuitively, we can imagine a line passing through these points that 'fits' the general pattern. For example, taking $m = 1.2$ and $c = 5$, the resulting line $y = 1.2x + 5$ could fit between the points we are given, as displayed in Figure 4.4.

This looks like an approximation of the relationship, but how do we know that it is the best approximation?

4.2.2　Minimising the error

For each of our observations, we can determine an error in the fitted model by calculating the difference between the real value of y and the one predicted

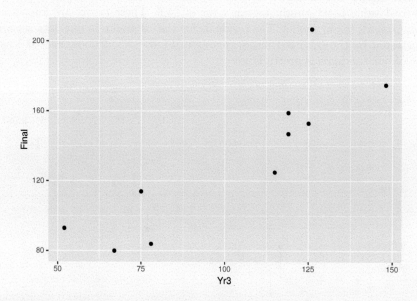

FIGURE 4.3: Basic scatter plot of 10 observations

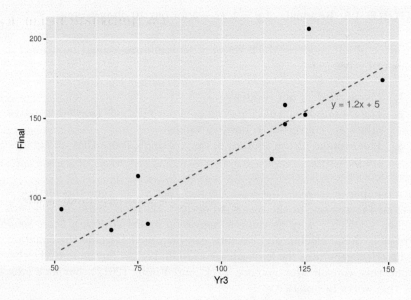

FIGURE 4.4: Fitting $y = 1.2x + 5$ to our 10 observations

by our model. For example, at $x = 52$, our modeled value of y is 67.4, but the real value is 93, producing an error of 25.6. These errors are known as the *residuals* of our model. The residuals for the 10 points in our data set are illustrated by the solid red line segments in Figure 4.5. It looks like at least one of our residuals is pretty large.

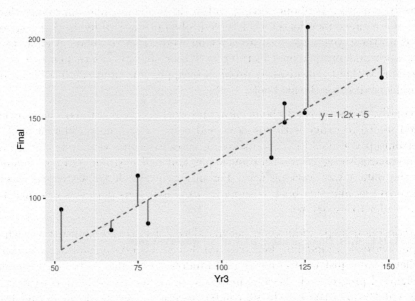

FIGURE 4.5: Residuals of $y = 1.2x + 5$ for our 10 observations

The error of our model—which we want to minimize—could be defined in a number of ways:

1. The average of our residuals
2. The average of the *absolute values* of our residuals (so that negative values are converted to positive values)
3. The average of the squares of our residuals (note that all squares are positive)

For a number of reasons (not least the fact that at the time this method was developed it was one of the easiest to derive), the most common approach is number 3, which is why we call our regression model *Ordinary Least Squares* regression. Some algebra and calculus can help us determine the equation of the line that generates the least-squared residual error. For more of the theory behind this, consult Montgomery, Peck, and Vining (2012), but let's look at how this works in practice.

4.2.3 Determining the best fit

We can run a fairly simple function in R to calculate the best fit linear model for our data. Once we have run that function, the model and all the details will be saved in our session for further investigation or use.

First we need to express the model we are looking to calculate as a formula. In this simple case, we want to regress the outcome y = Final against the input x = Yr3, and therefore we would use the simple formula notation Final ~ Yr3. Now we can use the lm() function to calculate the linear model based on our data set and our formula.

```
# calculate model
model <- lm(formula = Final ~ Yr3, data = d)
```

The model object that we have created is a list of a number of different pieces of information, which we can see by looking at the names of the objects in the list.

```
# view the names of the objects in the model in a column
model_objects <- names(model)
as.data.frame(model_objects)
```

```
##       model_objects
## 1      coefficients
## 2         residuals
## 3           effects
## 4              rank
## 5    fitted.values
## 6            assign
## 7                qr
## 8       df.residual
## 9           xlevels
## 10             call
## 11            terms
## 12            model
```

So we can already see some terms we are familiar with. For example, we can look at the coefficients.

`model$coefficients`

```
## (Intercept)        Yr3
##    16.630452   1.143257
```

This tells us that that our best fit model—the one that minimizes the average squares of the residuals—is $y = 1.14x + 16.63$. In other words, our Final test score can be expected to take a value of 16.63 even with zero score in the Yr3 input, and every additional point scored in Yr3 will increase the Final score by 1.14.

4.2.4 Measuring the fit of the model

We have calculated a model which minimizes the average squared residual error for the sample of data that we have, but we don't really have a sense of how 'good' the model is. How do we tell how well our model uses the input data to explain the outcome? This is an important question to answer because you would not want to propose a model that does not do a good job of explaining your outcome, and you also may need to compare your model to other alternatives, which will require some sort of benchmark metric. One natural way to benchmark how good a job your model does of explaining the outcome is to compare it to a situation where you have no input and no model at all. In this situation, all you have is your outcome values, which can be considered a random variable with a mean and a variance. In the case of our 10 observations, we have 10 values of Final with a mean of 133.7. We can consider the horizontal line representing the mean of y as our 'random model', and we can calculate the residuals around the mean. This can be seen in Figure 4.6.

Recall from Section 3.1.1 the definition of the population variance of y, notated as $\text{Var}(y)$. Note that it is defined as the average of the squares of the residuals around the mean of y. Therefore $\text{Var}(y)$ represents the average squared residual error of a random model. This calculates in this case to 1574.21. Let's overlay our fitted model onto this random model in Figure 4.7.

So for most of our observations (though not all) we seem to have reduced the 'distance' from the random model by fitting our new model. If we average the square of our residuals for the fitted model, we obtain the average squared residual error of our fitted model, which calculates to 398.35.

Therefore, before we fit our model, we have an error of 1574.21, and after we fit it, we have an error of 398.35. So we have reduced the error of our model by 1175.86 or, expressed as a proportion, by 0.75. In other words, we can say that our model explains 0.75 (or 75%) of the variance of our outcome.

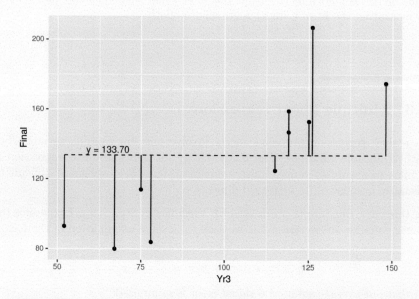

FIGURE 4.6: Residuals of our 10 observations around their mean value

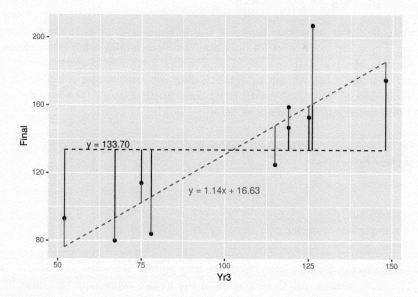

FIGURE 4.7: Comparison of residuals of fitted model (red) against random variable (blue)

This metric is known as the R^2 of our model and is the primary metric used in measuring the fit of a linear regression model[2].

4.3   Multiple linear regression

In reality, regression problems rarely involve one single input variable, but rather multiple variables. The methodology for multiple linear regression is similar in nature to simple linear regression, but obviously more difficult to visualize because of its increased dimensionality.

In this case, our inputs are a set of p variables $x_1, x_2, \dots, x_p$. Extending the linear equation in Section 4.2.1, we seek to develop an equation of the form:

$$y = \beta_0 + \beta_1 x_1 + \beta_2 x_2 + \cdots + \beta_p x_p$$

so that our average squared residual error is minimized.

4.3.1   Running a multiple linear regression model and interpreting its coefficients

A multiple linear regression model is run in a similar way to a simple linear regression model, with your formula notation determining what outcome and input variables you wish to have in your model. Let's now perform a multiple linear regression on our entire `ugtests` data set and regress our `Final` test score against all prior test scores using the formula `Final ~ Yr3 + Yr2 + Yr1` and determine our coefficients as before.

```
model <- lm(data = ugtests, formula = Final ~ Yr3 + Yr2 + Yr1)
model$coefficients
```

```
## (Intercept)         Yr3          Yr2          Yr1
## 14.14598945  0.86568123  0.43128539  0.07602621
```

Referring to our formula in Section 4.3, let's understand what each coefficient $\beta_0, \beta_1, \dots, \beta_p$ means. β_0, the *intercept* of the model, represents the value of y

---

[2]As a side note, in a simple regression model like this, where there is only one input variable, we have the simple identity $R^2 = r^2$, where r is the correlation between the input and outcome (for our small set of 10 observations here, the correlation is 0.864).

assuming that all the inputs were zero. You can imagine that your output can be expected to have a base value even without any inputs—a student who completely flunked the first three years can still redeem themselves to some extent in the Final year.

Now looking at the other coefficients, let's consider what happens if our first input x_1 increased by a single unit, assuming nothing else changed. We would then expect our value of y to increase by β_1. Similarly for any input x_k, a unit increase would result in an increase in y of β_k, assuming no other changes in the inputs.

In the case of our ugtests data set, we can say the following:

- The intercept of the model is 14.146. This is the value that a student could be expected to score on their final exam even if they had scored zero in all previous exams.

- The Yr3 coefficient is 0.866. Assuming no change in other inputs, this is the increase in the Final exam score that could be expected from an extra point in the Year 3 score.

- The Yr2 coefficient is 0.431. Assuming no change in other inputs, this is the increase in the Final exam score that could be expected from an extra point in the Year 2 score.

- The Yr1 coefficient is 0.076. Assuming no change in other inputs, this is the increase in the Final exam score that could be expected from an extra point in the Year 1 score.

4.3.2   Coefficient confidence

Intuitively, these coefficients appear too precise for comfort. After all, we are attempting to estimate a relationship based on a limited set of data. In particular, looking at the Yr1 coefficient, it seems to be very close to zero, implying that there is a possibility that the Year 1 examination score has no impact on the final examination score. Like in any statistical estimation, the coefficients calculated for our model have a margin of error. Typically, in any such situation, we seek to know a 95% confidence interval to set a standard of certainty around the values we are interpreting.

The summary() function is a useful way to gather critical information in your model, including important statistics on your coefficients:

```
model_summary <- summary(model)
model_summary$coefficients
```

```
##              Estimate Std. Error   t value       Pr(>|t|)
## (Intercept) 14.14598945 5.48005618  2.581358  9.986880e-03
## Yr3          0.86568123 0.02913754 29.710169 1.703293e-138
## Yr2          0.43128539 0.03250783 13.267124  4.860109e-37
## Yr1          0.07602621 0.06538163  1.162807  2.451936e-01
```

The 95% confidence interval corresponds to approximately two standard errors
above or below the estimated value. For a given coefficient, if this confidence
interval includes zero, you cannot reject the hypothesis that the variable has
no relationship with the outcome. Another indicator of this is the `Pr(>|t|)`
column of the coefficient summary, which represents the *p-value* of the null
hypothesis that the input variable has no relationship with the outcome. If
this value is less than a certain threshold (usually 0.05), you can conclude
that this variable has a statistically significant relationship with the outcome.
To see the precise confidence intervals for your model coefficients, you can use
the `confint()` function.

```
confint(model)
```

```
##                    2.5 %      97.5 %
## (Intercept)   3.39187185  24.9001071
## Yr3           0.80850142   0.9228610
## Yr2           0.36749170   0.4950791
## Yr1          -0.05227936   0.2043318
```

In this case, we can conclude that the examinations in Years 2 and 3 have
a significant relationship with the Final examination score, but we cannot
conclude this for Year 1. Effectively, this means that we can drop `Yr1` from
our model with no substantial loss of fit. In general, simpler models are easier
to manage and interpret, so let's remove the non-significant variable now.

```
newmodel <- lm(data = ugtests, formula = Final ~ Yr3 + Yr2)
```

Given that our new model only has three dimensions, we have the luxury of
visualizing it. Figure 4.8 shows the data and the fitted plane of our model.

4.3.3 Model 'goodness-of-fit'

At this point we can further explore the overall summary of our model. As you
saw in the previous section, our model summary contains numerous objects

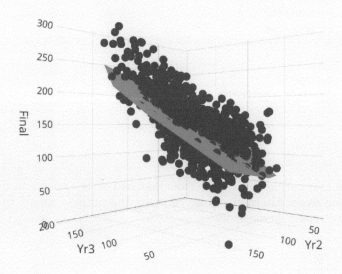

FIGURE 4.8: 3D visualization of the fitted `newmodel` against the `ugtests` data

of interest, including statistics on the coefficients of our model. We can see what is inside our summary by looking at the names of its contents, and we can then dive in and explore specific objects of interest.

```
# get summary of new model
newmodel_summary <- summary(newmodel)
```

```
# see summary contents
summary_objects <- names(newmodel_summary)
as.data.frame(summary_objects)
```

```
##      summary_objects
## 1              call
## 2             terms
## 3         residuals
## 4      coefficients
## 5           aliased
## 6             sigma
## 7                df
## 8         r.squared
## 9     adj.r.squared
```

```
## 10        fstatistic
## 11     cov.unscaled
```

```
# view r-squared
newmodel_summary$r.squared
```

```
## [1] 0.5296734
```

We can see that our model explains more than half of the variance in the Final examination score. Alternatively, we can view the entire summary to receive a formatted report on our model.

```
# see full model summary
newmodel_summary
```

```
##
## Call:
## lm(formula = Final ~ Yr3 + Yr2, data = ugtests)
##
## Residuals:
##     Min      1Q Median      3Q     Max
## -91.12 -20.36  -0.22   18.94   98.29
##
## Coefficients:
##              Estimate Std. Error t value Pr(>|t|)
## (Intercept) 18.08709    4.30701   4.199 2.92e-05 ***
## Yr3          0.86496    0.02914  29.687  < 2e-16 ***
## Yr2          0.43236    0.03250  13.303  < 2e-16 ***
## ---
## Signif. codes:  0 '***' 0.001 '**' 0.01 '*' 0.05 '.' 0.1 ' ' 1
##
## Residual standard error: 30.44 on 972 degrees of freedom
## Multiple R-squared:  0.5297, Adjusted R-squared:  0.5287
## F-statistic: 547.3 on 2 and 972 DF,  p-value: < 2.2e-16
```

This provides us with some of the most important metrics from our model. In particular, the last line gives us a report on our overall model confidence or 'goodness-of-fit'—this is a hypothesis test on the null hypothesis that our model does not fit the data any better than a random model. A high F-statistic indicates a strong likelihood that the model fits the data better than a random model. More intuitively, perhaps, we also have the p-value for the F-statistic. In this case it is extremely small, so we can reject the null hypothesis and conclude that our model has significant explanatory power over and above a random model.

Be careful not to confuse model goodness-of-fit with R^2. Depending on your sample, it is entirely possible for a model with a low R^2 to have high certainty for goodness-of-fit and vice versa.

4.3.4 Making predictions from your model

While this book focuses on inferential rather than predictive analytics, we briefly touch here on the mechanics of generating predictions from models. As you might imagine, once the model has been fitted, prediction is a relatively straightforward process. We feed the Yr2 and Yr3 examination scores into our fitted model, and it applies the coefficients to calculate the predicted outcome. Let's look at three fictitious students and create a dataframe with their scores to input into the model.

```
(new_students <- data.frame(
  Yr2 = c(67, 23, 88),
  Yr3 = c(144, 100, 166)
))
```

```
##    Yr2 Yr3
## 1   67 144
## 2   23 100
## 3   88 166
```

Now we can feed these values into our model to get predictions of the Final examination result for our three new students.

```
# use newmodel to predict for new_students
predict(newmodel, new_students)
```

```
##        1        2        3
## 171.6093 114.5273 199.7179
```

We know from our earlier work in this chapter that there is a confidence interval around the coefficients of our model, which means that there is a range

of values for our prediction according to those confidence intervals. This can
be determined by specifying that you require a confidence interval for your
predictions.

```
# get a confidence interval
predict(newmodel, new_students, interval = "confidence")
```

```
##          fit       lwr       upr
## 1  171.6093  168.2125  175.0061
## 2  114.5273  109.7081  119.3464
## 3  199.7179  195.7255  203.7104
```

You may also recall from Chapter 1 that any observation in our outcome is
subject to uncontrollable error, so that there is a further margin of 'prediction
error', even after we take into consideration the confidence interval of our
fitted model. Therefore to generate a more reliable prediction range to use in
real life, which takes this random, uncontrollable error into consideration, you
should calculate a 'prediction interval'.

```
# get a prediction interval
predict(newmodel, new_students, interval = "prediction")
```

```
##          fit        lwr        upr
## 1  171.6093  111.77795  231.4406
## 2  114.5273   54.59835  174.4562
## 3  199.7179  139.84982  259.5860
```

As discussed in Chapter 1, the process of developing a model to *predict* an
outcome can be quite different from developing a model to *explain* an outcome.
For a start, it is unlikely that you would use your entire sample to fit a
predictive model, as you would want to reserve a portion of your data to test
for its fit on new data. Since the focus of this book is inferential modeling,
much of this topic will be out of our scope.

4.4   Managing inputs in linear regression

Our walkthrough example for this chapter, while useful for illustrating the
key concepts, is a very straightforward data set to run a model on. There is

no missing data, and all the data inputs have the same numeric data type (in the exercises at the end of this chapter we will present a more varied data set for analysis). Commonly, an analyst will have a list of possible input variables that they can consider in their model, and rarely will they run a model using all of these variables. In this section we will cover some common elements of decision making and design of input variables in regression models.

4.4.1 Relevance of input variables

The first step in managing your input variables is to make a judgment about their relevance to the outcome being modeled. Analysts should not blindly run a model on a set of variables before considering their relevance. There are two common reasons for rejecting the inclusion of an input variable:

1. There is no reasonable possibility of a direct or indirect causal relationship between the input and the outcome. For example, if you were provided with the height of each individual taking the Final examination in our walkthrough example, it would be difficult to see how that could reasonably relate to the outcome that you are modeling.

2. If there is a possibility that the model will be used to predict based on new data in the future, there may be variables that you explicitly do not wish to be used in any prediction. For example, if our walkthrough model contained student gender data, we would not want to include that in a model that predicted future student scores because we would not want gender to be taken into consideration when determining student performance.

4.4.2 Sparseness ('missingness') of data

Missing data is a very common problem in modeling. If an observation has missing data in a variable that is being included in the model, that observation will be ignored, or an error will be thrown. This forces a model trained on a smaller set of data, which can compromise its powers of inference. Running summary functions on your data (such as `summary()` in R) will reveal variables that contain missing data if they exist.

There are three main options for how missing data is handled:

1. If the data for a given variable is relatively complete and only a small number of observations are missing, it's usually best and simplest to remove the observations that are missing from the data set. Note

that many modeling functions (though not all) will take care of this automatically.

2. As data becomes more sparse, removing observations becomes less of an option. If the sparseness is massive (for example, more than half of the data is missing), then there is no choice but to remove that variable from the model. While this may be unsatisfactory for a given variable (because it is thought to have an important explanatory role), the fact remains that data that is mostly missing is not a good measure of a construct in the first place.

3. Moderate sparse data could be considered for imputation. Imputation methods involve using the overall statistical properties of the entire data set or of specific other variables to 'suggest' what the missing value might be, ranging from simple mean and median values to more complex imputation methods. Imputation methods are more commonly used in predictive settings, and we will not cover imputation methods in depth here.

4.4.3   Transforming categorical inputs to dummy variables

Many models will have categorical inputs rather than numerical inputs. Categorical inputs usually take forms such as:

- Binary values—for example, Yes/No, True/False
- Unordered categories—for example Car, Train, Bicycle
- Ordered categories—for example Low, Medium, High

Categorical variables do not behave like numerical variables. There is no sense of quantity in a categorical variable. We do not know how a Car relates to a Train quantitatively, we only know that they are different. Even for an ordered category, although we know that 'Medium' is higher than 'Low', we do not know how much higher or indeed whether the difference is the same as that between 'High' and 'Medium'.

In general, all model input variables should take a numeric form. The most reliable way to do this is to convert categorical values to dummy variables. While some packages and functions have a built-in ability to convert categorical data to dummy variables, not all do, so it is important to know how to do this yourself. Consider the following data set:

```
(vehicle_data <- data.frame(
  make = c("Ford", "Toyota", "Audi"),
  manufacturing_cost = c(15000, 19000, 28000)
))
```

```
##      make manufacturing_cost
## 1    Ford               15000
## 2 Toyota               19000
## 3    Audi               28000
```

The make data is categorical, so it will be converted to several columns for each possible value of make, and binary labeling will be used to identify whether that value is present in that specific observation. Many packages and functions are available to conveniently do this, for example:

```
library(dummies)
(dummy_vehicle <- dummies::dummy("make", data = vehicle_data))
```

```
##    makeAudi makeFord makeToyota
## 1         0        1          0
## 2         0        0          1
## 3         1        0          0
```

Dummy variables can then replace the original make column to get your data set ready for modeling.

```
(vehicle_data_dummies <- cbind(
  manufacturing_cost = vehicle_data$manufacturing_cost,
  dummy_vehicle
))
```

```
##    manufacturing_cost makeAudi makeFord makeToyota
## 1               15000        0        1          0
## 2               19000        0        0          1
## 3               28000        1        0          0
```

It is worth a moment to consider how to interpret coefficients of dummy variables in a linear regression model. Note that all observations will have one of the dummy variable values (all cars must have a make). Therefore the model will assume a 'reference value' for the categorical variable—often this is the first value in alphabetical or numerical order. In this case, Audi would be the reference dummy variable. The model then calculates the effect on the outcome variable of a 'switch' from Audi to one of the other dummies[3]. If we were to try to use the data in our vehicle_data_dummies data set to explain the retail price of a vehicle, we would interpret coefficients like this:

---

[3]For more on how to control which categorical value is used as a reference, see Section 6.3.1.

- Comparing two cars of the same make, we would expect each *extra dollar* spent on manufacturing to change the retail price by ...
- Comparing a Ford with an Audi *of the same manufacturing cost*, we would expect a difference in retail price of ...
- Comparing a Toyota with an Audi *of the same manufacturing cost*, we would expect a difference in retail price of ...

This highlights the importance of appropriate interpretation of coefficients, and in particular the proper understanding of units. It will be common to see much larger coefficients for dummy variables in regression models because they represent a binary 'all' or 'nothing' variable in the model. The coefficient for manufacturing cost would be much smaller because a unit in this case is a dollar of manufacturing spend, on a scale of many thousands of potential dollars in spend. Care should be taken not to 'rank' coefficients by their value. Higher coefficients in and of themselves do not imply greater importance[4].

4.5   Testing your model assumptions

All modeling techniques have underlying assumptions about the data that they model and can generate inaccurate results when those assumptions do not hold true. Conscientious analysts will verify that these assumptions are satisfied before finalizing their modeling efforts. In this section we will outline some common checks of model assumptions when running linear regression models.

4.5.1   Assumption of linearity and additivity

Linear regression assumes that the relationship we are trying to model is linear and additive in nature. Therefore you can expect problems if you are using this approach to model a pattern that is not linear or additive.

You can check whether your linearity assumption was reasonable in a couple of ways. You can plot the true versus the predicted (fitted) values to see if they look correlated. You can see such a plot on our student examination model in Figure 4.9.

---

[4]Rescaling numerical input variables onto common scales can help with understanding the ranked importance of these variables. In some techniques, for example structural modeling which we will review in Section 8.2, scaled regression coefficients help determine the ranked importance of constructs to the outcome.

```
predicted_values <- newmodel$fitted.values
true_values <- ugtests$Final

# plot true values against predicted values
plot(predicted_values, true_values)
```

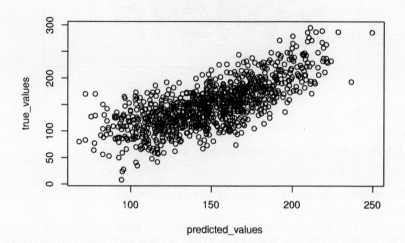

FIGURE 4.9: Plot of true versus fitted/predicted student scores

Alternatively, you can plot the residuals of your model against the predicted values and look for the pattern of a random distribution (that is, no major discernible pattern) such as in Figure 4.10.

```
residuals <- newmodel$residuals

# plot residuals against predicted values
plot(predicted_values, residuals)
```

You can also plot the residuals against each input variable as an extra check of independent randomness, looking for a reasonably random distribution in all cases. If you find that your residuals are following a clear pattern and are not random in nature, this is an indication that a linear model is not a good choice for your data.

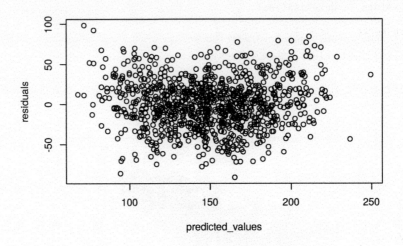

FIGURE 4.10: Plot of residuals against fitted/predicted scores

4.5.2 Assumption of constant error variance

It is assumed in a linear model that the errors or residuals are *homoscedastic*—this means that their variance is constant across the values of the input variables. If the errors of your model are *heteroscedastic*—that is, if they increase or decrease according to the value of the model inputs—this can lead to poor estimations and inaccurate inferences.

While a simple plot of residuals against predicted values (such as in Figure 4.10) can give a quick indication on homoscedacity, to be thorough the residuals should be plotted against each input variable, and it should be verified that the range of the residuals remains broadly stable. In our student examination model, we can first plot the residuals against the values of Yr2 in Figure 4.11.

```
Yr2 <- ugtests$Yr2

# plot residuals against Yr2 values
plot(Yr2, residuals)
```

We see a pretty consistent range of values for the residuals in 4.11. Similarly we can plot the residuals against the values of Yr3, as in Figure 4.12.

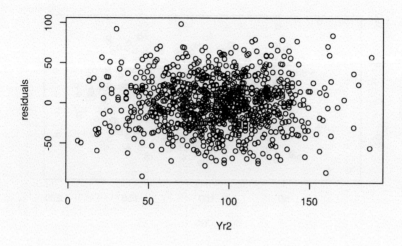

FIGURE 4.11: Plot of residuals against `Yr2` values

```
Yr3 <- ugtests$Yr3

# plot residuals against Yr3 values
plot(Yr3, residuals)
```

Figure 4.12 also shows a consistent range of values for the residuals, which reassures us that we have homoscedacity.

4.5.3 Assumption of normally distributed errors

In an appropriate model we expect our errors to be random, so we would therefore expect our residuals to be normally distributed over sufficient numbers of observations. If our residuals are distributed differently, this is again an indicator of an inappropriate model and can result in inaccurate estimates of confidence intervals and the statistical significance of coefficients.

The quickest way to determine if residuals in your sample are consistent with a normal distribution is to run a quantile-quantile plot (or Q-Q plot) on the residuals. This will plot the observed quantiles of your sample against the theoretical quantiles of a normal distribution. The closer this plot looks like a perfect correlation, the more certain you can be that this normality as-

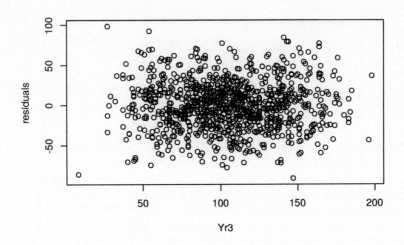

FIGURE 4.12: Plot of residuals against `Yr3` values

sumption holds. An example for our student examination model is in Figure 4.13.

```
# normal distribution qqplot of residuals
qqnorm(newmodel$residuals)
```

4.5.4 Avoiding high collinearity and multicollinearity between input variables

In multiple linear regression, the various input variables used can be considered 'dimensions' of the problem or model. In theory, we ideally expect dimensions to be independent and uncorrelated. Practically speaking, however, it's very challenging in large data sets to ensure that every input variable is completely uncorrelated from another. For example, even in our limited `ugtests` data set we saw in Figure 4.2 that `Yr2` and `Yr3` examination scores are correlated to some degree.

While some correlation between input variables can be expected and tolerated in linear regression models, high levels of correlation can result in significant inflation of coefficients and inaccurate estimates of p-values of coefficients.

Collinearity means that two input variables are highly correlated. The definition of 'high correlation' is a matter of judgment, but as a rule of thumb

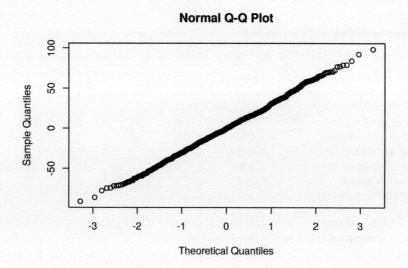

FIGURE 4.13: Quantile-quantile plot of residuals

correlations greater than 0.5 might be considered high and greater than 0.7 might be considered extreme. Creating a simple correlation matrix or a pair-plot (such as Figure 4.2) can immediately surface high or extreme collinearity.

Multicollinearity means that there is a linear relationship between more than two of the input variables. This may not always present itself in the form of high correlations between pairs of input variables, but may be seen by identifying 'clusters' of moderately correlated variables, or by calculating a Variance Inflation Factor (VIF) for each input variable—where VIFs greater than 5 indicate high multicollinearity. Easy-to-use tests also exist in statistical software for identifying multicollinearity (for example the mctest package in R). Here is how we would test for multicollinearity in our student examination model.

```
library(mctest)

# diagnose possible overall presence of multicollinearity
mctest::omcdiag(newmodel)
```

```
##
## Call:
## mctest::omcdiag(mod = newmodel)
##
##
## Overall Multicollinearity Diagnostics
##
##                            MC Results detection
## Determinant |X'X|:           0.9981          0
## Farrar Chi-Square:           1.8365          0
## Red Indicator:               0.0434          0
## Sum of Lambda Inverse:       2.0038          0
## Theil's Method:             -0.5259          0
## Condition Number:            9.1952          0
##
## 1 --> COLLINEARITY is detected by the test
## 0 --> COLLINEARITY is not detected by the test
```

```
# if necessary, diagnose specific multicollinear variables using VIF
mctest::imcdiag(newmodel, method = "VIF")
```

```
##
## Call:
## mctest::imcdiag(mod = newmodel, method = "VIF")
##
##
##   VIF Multicollinearity Diagnostics
##
##         VIF detection
## Yr3 1.0019         0
## Yr2 1.0019         0
##
## NOTE:  VIF Method Failed to detect multicollinearity
##
##
## 0 --> COLLINEARITY is not detected by the test
##
## ====================================
```

Note that collinearity and multicollinearity only affect the coefficients of the variables impacted, and do not affect other variables or the overall statistics and fit of a model. Therefore, if a model is being developed primarily to make predictions and there is little interest in using the model to explain a phenomenon, there may not be any need to address this issue at all. However, in inferential modeling the accuracy of the coefficients is very important, and so testing of multicollinearity is essential. In general, the best way to deal with collinear variables is to remove one of them from the model (usually the one that has the least significance in explaining the outcome).

4.6 Extending multiple linear regression

We wrap up this chapter by introducing some simple extensions of linear regression, with a particular aim of trying to improve the overall fit of a model by relaxing the linear or additive assumptions. It is rare for practitioners to extend linear regression models too greatly due to the negative impact this can have on interpretation, but simple extensions such as experimenting with interaction terms or quadratics are not uncommon. If you have an appetite to explore this topic more fully, I recommend Rao et al. (2008).

4.6.1 Interactions between input variables

Recall that our model of student examination scores took each year's score as an independent input variable, and therefore we are making the assumption that the score obtained in each year acts independently and additively in predicting the Final score. However, it is very possible that several input variables act together in relation to the outcome. One way of modeling this is to include interaction terms in your model, which are new input variables formed as products of the original input variables.

In our student examination data in ugtests, we could consider extending our model to not only include the individual year examinations, but also to include the impact of combined changes across multiple years. For example, we could combine the impact of Yr2 and Yr3 examinations by multiplying them together in our model.

```
interaction_model <- lm(data = ugtests,
                        formula = Final ~ Yr2 + Yr3 + Yr2*Yr3)
summary(interaction_model)
```

```
##
## Call:
## lm(formula = Final ~ Yr2 + Yr3 + Yr2 * Yr3, data = ugtests)
##
## Residuals:
##      Min       1Q   Median       3Q      Max
## -78.084  -18.284   -0.546   18.395   79.824
##
## Coefficients:
##                Estimate Std. Error t value Pr(>|t|)
## (Intercept)   1.320e+02  1.021e+01  12.928  < 2e-16 ***
## Yr2          -7.947e-01  1.056e-01  -7.528 1.18e-13 ***
## Yr3          -2.267e-01  9.397e-02  -2.412   0.0161 *
## Yr2:Yr3       1.171e-02  9.651e-04  12.134  < 2e-16 ***
## ---
## Signif. codes:  0 '***' 0.001 '**' 0.01 '*' 0.05 '.' 0.1 ' ' 1
##
## Residual standard error: 28.38 on 971 degrees of freedom
## Multiple R-squared:  0.5916, Adjusted R-squared:  0.5903
## F-statistic: 468.9 on 3 and 971 DF,  p-value: < 2.2e-16
```

We see that introducing this interaction term has improved the fit of our model from 0.53 to 0.59, and that the interaction term is significant, so we conclude that in addition to a significant effect of the Yr2 and Yr3 scores, there is an additional significant effect from their interaction Yr2*Yr3. Let's take a moment to understand how to interpret this, since we note that some of the coefficients are now negative.

Our model now includes two input variables and their interaction, so it can be written as

$$
\begin{aligned}
\text{Final} &= \beta_0 + \beta_1 \text{Yr3} + \beta_2 \text{Yr2} + \beta_3 \text{Yr3Yr2} \\
&= \beta_0 + (\beta_1 + \beta_3 \text{Yr2})\text{Yr3} + \beta_2 \text{Yr2} \\
&= \beta_0 + \gamma \text{Yr3} + \beta_2 \text{Yr2}
\end{aligned}
$$

where $\gamma = \beta_1 + \beta_3 \text{Yr2}$. Therefore our model has coefficients which are not constant but change with the values of the input variables. We can conclude that the effect of an extra point in the examination in Year 3 will be different depending on how the student performed in Year 2. Visualizing this, we can see in Figure 4.14 that this non-constant term introduces a curvature to our fitted surface that aligns it a little more closely with the observations in our data set.

By examining the shape of this curved plane, we can observe that the model considers *trajectories* in the Year 2 and Year 3 examination scores. Those individuals who have improved from one year to the next will perform better

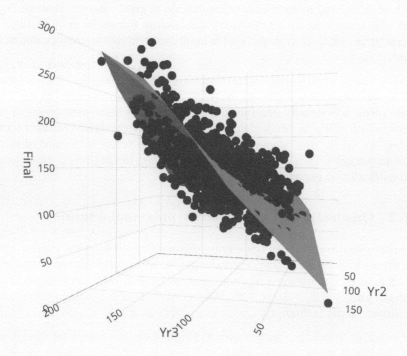

FIGURE 4.14: 3D visualization of the fitted `interaction_model` against the `ugtests` data

in this model than those who declined. To demonstrate, let's look at the predicted scores from our `interaction_model` for someone who declined and for someone who improved from Year 2 to Year 3.

```
# data frame with a declining and an improving observation
obs <- data.frame(
  Yr2 = c(150, 75),
  Yr3 = c(75, 150)
)

predict(interaction_model, obs)
```

```
##        1        2
## 127.5010 170.1047
```

Through including the interaction effect, the model interprets declining examination scores more negatively than improving examination scores. These kinds of additional inferential insights may be of great interest. However, consider the impact on interpretability of modeling too many combinations of interactions. As always, there is a trade-off between intepretability and accuracy[5].

When running models with interaction terms, you can expect to see a hierarchy in the coefficients according to the level of the interaction. For example, single terms will usually generate higher coefficients than interactions of two terms, which will generate higher coefficients than interactions of three terms, and so on. Given this, whenever an interaction of terms is considered significant in a model, then the single terms contained in that interaction should automatically be regarded as significant.

4.6.2 Quadratic and higher-order polynomial terms

In many situations the real underlying relationship between the outcome and the inputs may be non-linear. For example, if the underlying relationship was thought to be quadratic on a given input variable x, then the formula would take the form $y = \beta_0 + \beta_1 x + \beta_2 x^2$. We can easily trial polynomial terms using our linear model technology.

For example, recall that we removed Yr1 data from our model because it was not significant when modeled linearly. We could test if a quadratic model on Yr1 helps improve our fit[6]:

```
# add a quadratic term in Yr1
quadratic_yr1_model <- lm(data = ugtests,
                          formula = Final ~ Yr3 + Yr2 + Yr1 + I(Yr1^2))
```

```
# test R-squared
summary(quadratic_yr1_model)$r.squared
```

```
## [1] 0.5304198
```

In this case we find that modeling Yr1 as a quadratic makes no difference to the fit of the model.

---

[5]In a predictive context, there is also the issue of 'overfitting' the model, where the model is too 'tightly' aligned to the past data that was used in fitting it that it may be very inaccurate for new data. For example, in our interaction model, someone who scores very low in both Year 2 and Year 3 will be awarded an unreasonably high score (see the intercept coefficient in the interactive model summary). This reinforces the need to test and validate model fits in a predictive context.

[6]Note the use of I() in the formula notation here. This is because the symbol ^ has a different meaning inside a formula, and we use I() to *isolate* what is inside the parentheses to ensure that it is interpreted literally as 'the square of Yr1'.

4.7 Learning exercises

4.7.1 Discussion questions

1. What is the approximate meaning of the term 'regression'? Why is the term particularly suited to the methodology described in this chapter?

2. What basic condition must the outcome variable satisfy for linear regression to be a potential modeling approach? Describe some ideas for problems that might be modeled using linear regression.

3. What is the difference between simple linear regression and multiple linear regression?

4. What is a residual, and how does it relate to the term 'Ordinary Least Squares'?

5. How are the coefficients of a linear regression model interpreted? Explain why higher coefficients do not necessarily imply greater importance.

6. How is the R^2 of a linear regression model interpreted? What are the minimum and maximum possible values for R^2, and what does each mean?

7. What are the key considerations when preparing input data for a linear regression model?

8. Describe your understanding of the term 'dummy variable'. Why are dummy variable coefficients often larger than other coefficients in linear regression models?

9. Describe the term 'collinearity' and why it is an important consideration in regression models.

10. Describe some ways that linear regression models can be extended into non-linear models.

4.7.2 Data exercises

Load the sociological_data data set via the peopleanalyticsdata package or download it from the internet[7]. This data represents a sample of information obtained from individuals who participated in a global research study and contains the following fields:

---

[7]http://peopleanalytics-regression-book.org/data/sociological_data.csv

- `annual_income_ppp`: The annual income of the individual in PPP adjusted US dollars
- `average_wk_hrs`: The average number of hours per week worked by the individual
- `education_months`: The total number of months spent by the individual in formal primary, secondary and tertiary education
- `region`: The region of the world where the individual lives
- `job_type`: Whether the individual works in a skilled or unskilled profession
- `gender`: The gender of the individual
- `family_size`: The size of the individual's family of dependents
- `work_distance`: The distance between the individual's residence and workplace in kilometers
- `languages`: The number of languages spoken fluently by the individual

Conduct some exploratory data analysis on this data set. Including:

1. Identify the extent to which missing data is an issue.
2. Determine if the data types are appropriate for analysis.
3. Using a correlation matrix, pairplot or alternative method, identify whether collinearity is present in the data.
4. Identify and discuss anything else interesting that you see in the data.

Prepare to build a linear regression model to explain the variation in `annual_income_ppp` using the other data in the data set.

5. Are there any fields which you believe should not be included in the model? If so, why?
6. Would you consider imputing missing data for some or all fields where it is an issue? If so, what might be some simple ways to impute the missing data?
7. Which variables are categorical? Convert these variables to dummy variables using a convenient function or using your own approach.

Run and interpret the model. For convenience, and to avoid long formula strings, you can use the formula notation `annual_income_ppp ~ .` which means 'regress `annual_income` against everything else'. You can also remove fields this way, for example `annual_income_ppp ~ . - family_size`.

8. Determine what variables are significant predictors of annual income and what is the effect of each on the outcome.
9. Determine the overall fit of the model.
10. Do some simple analysis on the residuals of the model to determine if the model is safe to interpret.

11. Experiment with improving the model fit through possible interaction terms or non-linear extensions.
12. Comment on your results. Did anything in the results surprise you? If so, what might be possible explanations for this.
13. Explain why you would or would not be comfortable using a model like this in a predictive setting—for example to help employers determine the right pay for employees.

5

Binomial Logistic Regression for Binary Outcomes

In the previous chapter we looked at how to explain outcomes that have continuous scale, such as quantity, money, height or weight. While there are a number of typical outcomes of this type in the people analytics domain, they are not the most common form of outcomes that are typically modeled. Much more common are situations where the outcome of interest takes the form of a limited set of classes. Binary (two class) problems are very common. Hiring, promotion and attrition are often modeled as binary outcomes: for example 'Promoted' or 'Not promoted.' Multi-class outcomes like performance ratings on an ordinal scale, or survey responses on a Likert scale are often converted to binary outcomes by dividing the ratings into two groups, for example 'High' and 'Not High.'

In any situation where our outcome is binary, we are effectively working with likelihoods. These are not generally linear in nature, and so we no longer have the comfort of our inputs being *directly* linearly related to our outcome. Therefore direct linear regression methods such as Ordinary Least Squares regression are not well suited to outcomes of this type. Instead, linear relationships can be inferred on *transformations* of the outcome variable, which gives us a path to building interpretable models. Hence, binomial logistic regression is said to be in a class of *generalized linear models* or *GLMs*. Understanding logistic regression and using it reliably in practice is not straightforward, but it is an invaluable skill to have in the people analytics domain. The mathematics of this chapter is a little more involved but worth the time investment in order to build a competent understanding of how to interpret these types of models.

DOI: 10.1201/9781003194156-5

5.1   When to use it

5.1.1   Origins and intuition of binomial logistic regression

The *logistic function* was first introduced by the Belgian mathematician Pierre François Verhulst in the mid-1800s as a tool for modeling population growth for humans, animals and certain species of plants and fruits. By this time, it was generally accepted that population growth could not continue exponentially forever, and that there were environmental and resource limits which place a maximum limit on the size of a population. The formula for Verhulst's function was:

$$y = \frac{L}{1 + e^{-k(x - x_0)}}$$

where e is the exponential constant, x_0 is the value of x at the midpoint, L is the maximum value of y (known as the 'carrying capacity') and k is the maximum gradient of the curve.

The logistic function, as shown in Figure 5.1, was felt to accurately capture the theorized stages of population growth, with slower growth in the initial stage, moving to exponential growth during the intermediate stage and then to slower growth as the population approaches its carrying capacity.

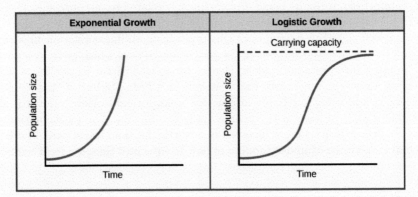

FIGURE 5.1: Verhulst's logistic function modeled both the exponential nature and the natural limit of population growth

In the early 20th century, starting with applications in economics and in chemistry, the logistic function was adopted in a wide array of fields as a useful tool for modeling phenomena. In statistics, it was observed that the logistic function has a similar S-shape (or *sigmoid*) to a cumulative normal distribution of probability, as depicted in Figure 5.2[1], where the x scale represents standard deviations around a mean. As we will learn, the logistic function gives rise to a mathematical model where the coefficients are easily interpreted in terms of likelihood of the outcome. Unsurprisingly, therefore, the logistic model soon became a common approach to modeling probabilistic phenomena.

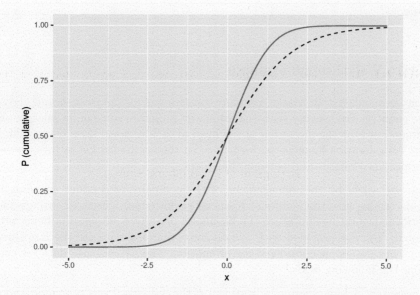

FIGURE 5.2: The logistic function (blue dashed line) is very similar to a cumulative normal distribution (red solid line) but easier to interpret

5.1.2 Use cases for binomial logistic regression

Binomial logistic regression can be used when the outcome of interest is binary or dichotomous in nature. That is, it takes one of two values. For example, one or zero, true or false, yes or no. These classes are commonly described as 'positive' and 'negative' classes. There is an underlying assumption that the cumulative probability of the outcome takes a shape similar to a cumulative normal distribution.

Here are some example questions that could be approached using binomial logistic regression:

- Given a set of data about sales managers in an organization, including performance against targets, team size, tenure in the organization and other

---

[1]The logistic function plotted in Figure 5.2 takes the simple form $y = \frac{1}{1+e^{-x}}$.

factors, what influence do these factors have on the likelihood of the individual receiving a high performance rating?

- Given a set of demographic, income and location data, what influence does each have on the likelihood of an individual voting in an election?
- Given a set of statistics about the in-game activity of soccer players, what relationship does each statistic have with the likelihood of a player scoring a goal?

5.1.3 Walkthrough example

You are an analyst for a large company consisting of regional sales teams across the country. Twice every year, this company promotes some of its salespeople. Promotion is at the discretion of the head of each regional sales team, taking into consideration financial performance, customer satisfaction ratings, recent performance ratings and personal judgment.

You are asked by the management of the company to conduct an analysis to determine how the factors of financial performance, customer ratings and performance ratings influence the likelihood of a given salesperson being promoted. You are provided with a data set containing data for the last three years of salespeople considered for promotion. The salespeople data set contains the following fields:

- promoted: A binary value indicating 1 if the individual was promoted and 0 if not
- sales: the sales (in thousands of dollars) attributed to the individual in the period of the promotion
- customer_rate: the average satisfaction rating from a survey of the individual's customers during the promotion period
- performance: the most recent performance rating prior to promotion, from 1 (lowest) to 4 (highest)

Let's take a quick look at the data.

```
# if needed, download salespeople data
url <- "http://peopleanalytics-regression-book.org/data/salespeople.csv"
salespeople <- read.csv(url)
```

```
# look at the first few rows of data
head(salespeople)
```

```
##   promoted sales customer_rate performance
## 1        0   594          3.94           2
## 2        0   446          4.06           3
## 3        1   674          3.83           4
## 4        0   525          3.62           2
## 5        1   657          4.40           3
## 6        1   918          4.54           2
```

The data looks as expected. Let's get a summary of the data.

```
summary(salespeople)
```

```
##      promoted            sales        customer_rate     performance
##   Min.   :0.0000   Min.   :151.0   Min.   :1.000   Min.   :1.0
##   1st Qu.:0.0000   1st Qu.:389.2   1st Qu.:3.000   1st Qu.:2.0
##   Median :0.0000   Median :475.0   Median :3.620   Median :3.0
##   Mean   :0.3219   Mean   :527.0   Mean   :3.608   Mean   :2.5
##   3rd Qu.:1.0000   3rd Qu.:667.2   3rd Qu.:4.290   3rd Qu.:3.0
##   Max.   :1.0000   Max.   :945.0   Max.   :5.000   Max.   :4.0
##                    NA's   :1       NA's   :1       NA's   :1
```

First we see a small number of missing values, and we should remove those observations. We see that about a third of individuals were promoted, that sales ranged from $151k to $945k, that as expected the average satisfaction ratings range from 1 to 5, and finally we see four performance ratings, although the performance categories are numeric when they should be an ordered factor, and promoted is numeric when it should be categorical. Let's convert these, and then let's do a pairplot to get a quick view on some possible underlying relationships, as in Figure 5.3.

```
library(GGally)

# remove NAs
salespeople <- salespeople[complete.cases(salespeople), ]

# convert performance to ordered factor and promoted to categorical
salespeople$performance <- ordered(salespeople$performance,
                                   levels = 1:4)
salespeople$promoted <- as.factor(salespeople$promoted)

# generate pairplot
GGally::ggpairs(salespeople)
```

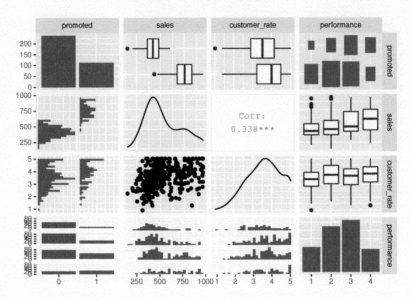

FIGURE 5.3: Pairplot for the salespeople data set

We can see from this pairplot that there are clearly higher sales for those who are promoted versus those who are not. We also see a moderate relationship between customer rating and sales, which is intuitive (if the customer doesn't think much of you, sales wouldn't likely be very high).

So we can see that some relationships with our outcome may exist here, but it's not clear how to tease them out and quantify them relative to each other. Let's explore how binomial logistic regression can help us do this.

5.2   Modeling probabilistic outcomes using a logistic function

Imagine that you have an outcome event y which either occurs or does not occur. The probability of y occurring, or $P(y = 1)$, obviously takes a value between 0 and 1. Now imagine that some input variable x has a positive effect on the probability of the event occurring. Then you would naturally expect $P(y = 1)$ to increase as x increases.

In our salespeople data set, let's plot our promotion outcome against the sales input. This can be seen in Figure 5.4.

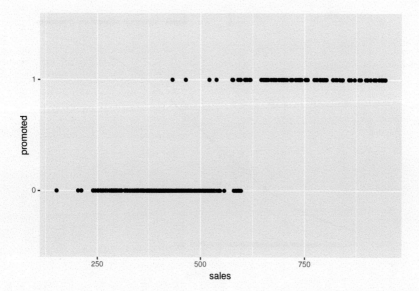

FIGURE 5.4: Plot of promotion against sales in the `salespeople` data set

It's clear that promotion is more likely with higher sales levels. As we move along the x axis from left to right and gradually include more and more individuals with higher sales, we know that the probability of promotion is gradually increasing overall. We could try to model this probability using our logistic function, which we learned about in Section 5.1.1. For example, let's plot the logistic function

$$P(y = 1) = \frac{1}{1 + e^{-k(x-x_0)}}$$

on this data, where we set x_0 to the mean of `sales` and k to be some maximum gradient value. In Figure 5.5 we can see these logistic functions for different values of k. All of these seem to reflect the pattern we are observing to some extent, but how do we determine the best-fitting logistic function?

5.2.1 Deriving the concept of log odds

Let's look more carefully at the index of the exponential constant e in the denominator of our logistic function. Note that, because x_0 is a constant, we have:

$$-k(x - x_0) = -(-kx_0 + kx) = -(\beta_0 + \beta_1 x)$$

where $\beta_0 = -kx_0$ and $\beta_1 = k$. Therefore,

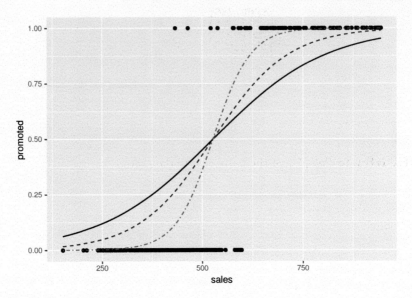

FIGURE 5.5: Overlaying logistic functions with various gradients onto previous plot

$$P(y = 1) = \frac{1}{1 + e^{-(\beta_0 + \beta_1 x)}}$$

This equation makes intuitive sense. As the value of x increases, the value $e^{-(\beta_0 + \beta_1 x)}$ gets smaller and smaller towards zero, and thus $P(y = 1)$ approaches its theoretical maximum value of 1. As the value of x decreases towards zero, we see that the value of $P(y = 1)$ approaches a minimum value of $\frac{1}{1 + e^{-\beta_0}}$. Referring back to our salespeople example, we can thus see that β_0 helps determine the baseline probability of promotion assuming no sales at all. If β_0 has an extremely negative value, this baseline probability will approach its theoretical minimum of zero.

Let's formalize the role of β_0 and β_1 in the likelihood of a positive outcome. We know that for any binary event y, $P(y = 0)$ is equal to $1 - P(y = 1)$, so

$$\begin{aligned} P(y = 0) &= 1 - \frac{1}{1 + e^{-(\beta_0 + \beta_1 x)}} \\ &= \frac{1 + e^{-(\beta_0 + \beta_1 x)} - 1}{1 + e^{-(\beta_0 + \beta_1 x)}} \\ &= \frac{e^{-(\beta_0 + \beta_1 x)}}{1 + e^{-(\beta_0 + \beta_1 x)}} \end{aligned}$$

Putting these together, we find that

$$\frac{P(y=1)}{P(y=0)} = \frac{\frac{1}{1+e^{-(\beta_0+\beta_1 x)}}}{\frac{e^{-(\beta_0+\beta_1 x)}}{1+e^{-(\beta_0+\beta_1 x)}}}$$

$$= \frac{1}{e^{-(\beta_0+\beta_1 x)}}$$

$$= e^{\beta_0 + \beta_1 x}$$

or alternatively, if we apply the natural logarithm to both sides

$$\ln\left(\frac{P(y=1)}{P(y=0)}\right) = \beta_0 + \beta_1 x$$

The right-hand side should look familiar from the previous chapter on linear regression, meaning there is something here we can model linearly. But what is the left-hand side?

$P(y=1)$ is the probability that the event will occur, while $P(y=0)$ is the probability that the event will not occur. You may be familiar from sports like horse racing or other gambling situations that the ratio of these two represents the *odds* of an event. For example, if a given horse has odds of 1:4, this means that there is a 20% probability they will win and an 80% probability they will not[2].

Therefore we can conclude that the natural logarithm of the odds of y—usually termed the *log odds* of y—is linear in x, and therefore we can model the log odds of y using similar linear regression methods to those studied in Chapter 4[3].

5.2.2 Modeling the log odds and interpreting the coefficients

Let's take our simple case of regressing the promoted outcome against sales. We use a standard binomial GLM function and our standard formula notation which we learned in the previous chapter.

```
# run a binomial model
sales_model <- glm(formula = promoted ~ sales,
                   data = salespeople, family = "binomial")

# view the coefficients
sales_model$coefficients
```

---

[2] Often in sports the odds are expressed in the reverse order, but the concept is the same.
[3] In this case, a more general form of the Ordinary Least Squares procedure is used to fit the model, known as *maximum likelihood estimation*.

```
##   (Intercept)          sales
## -21.77642020    0.03675848
```

We can interpret the coefficients as follows:

1. The (Intercept) coefficient is the value of the log odds with zero input value of x—it is the log odds of promotion if you made no sales.

2. The sales coefficient represents the increase in the log odds of promotion associated with each unit increase in sales.

We can convert these coefficients from log odds to odds by applying the exponent function, to return to the identity we had previously

$$\frac{P(y=1)}{P(y=0)} = e^{\beta_0+\beta_1 x} = e^{\beta_0}(e^{\beta_1})^x$$

From this, we can interpret that e^{β_0} represents the base odds of promotion assuming no sales, and that for every additional unit sales, those base odds are multiplied by e^{β_1}. Given this multiplicative effect that e^{β_1} has on the odds, it is known as an *odds ratio*.

```
# convert log odds to base odds and odds ratio
exp(sales_model$coefficients)
```

```
##   (Intercept)          sales
## 3.488357e-10 1.037442e+00
```

So we can see that the base odds of promotion with zero sales is very close to zero, which makes sense. Note that odds can only be precisely zero in a situation where it is impossible to be in the positive class (that is, nobody gets promoted). We can also see that each unit (that is, every $1000) of sales multiplies the base odds by approximately 1.04—in other words, it increases the odds of promotion by 4%.

5.2.3   Odds versus probability

It is worth spending a little time understanding the concept of odds and how it relates to probability. It is extremely common for these two terms to be used synonymously, and this can lead to serious misunderstandings when interpreting a logistic regression model.

If a certain event has a probability of 0.1, then this means that its odds are 1:9, or 0.111. If the probability is 0.5, then the odds are 1, if the probability is 0.9, then the odds are 9, and if the probability is 0.99, the odds are 99. As we approach a probability of 1, the odds become exponentially large, as illustrated in Figure 5.6:

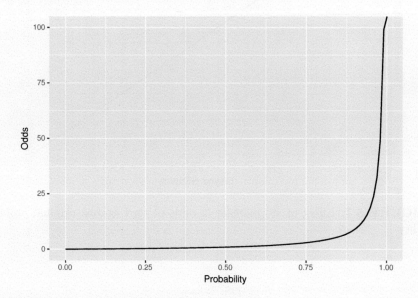

FIGURE 5.6: Odds plotted against probability

The consequence of this is that a given increase in odds can have a different effect on probability depending on what the original probability was in the first place. If the probability was already quite low, for example 0.1, then a 4% increase in odds translates to odds of 0.116, which translates to a new probability of 0.103586, representing an increase in probability of 3.59%, which is very close to the increase in odds. If the probability was already high, say 0.9, then a 4% increase in odds translates to odds of 9.36, which translates to a new probability of 0.903475 representing an increase in probability of 0.39%, which is very different from the increase in odds. Figure 5.7 shows the impact of a 4% increase in odds according to the original probability of the event.

We can see that the closer the base probability is to zero, the similar the effect of the increase on both odds and on probability. However, the higher the probability of the event, the less impact the increase in odds has. In any case, it's useful to remember the formulas for converting odds to probability and vice versa. If O represents odds and P represents probability then we have:

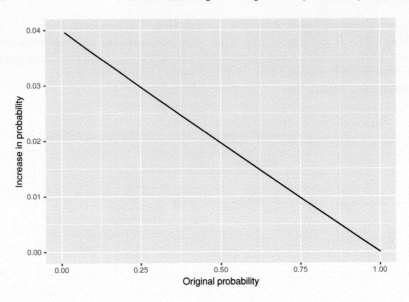

FIGURE 5.7: Effect of 4% increase in odds plotted against original probability

$$O = \frac{P}{1 - P}$$

$$P = \frac{O}{1 + O}$$

5.3 Running a multivariate binomial logistic regression model

The derivations in the previous section extend to multivariate data. Let y be a dichotomous outcome, and let $x_1, x_2, \ldots, x_p$ be our input variables. Then

$$\ln\left(\frac{P(y=1)}{P(y=0)}\right) = \beta_0 + \beta_1 x_1 + \beta_2 x_2 + \cdots + \beta_p x_p$$

for coefficients $\beta_0, \beta_1, \ldots, \beta_p$. As before:

- β_0 represents the log odds of our outcome when all inputs are zero
- Each β_i represents the increase in the log odds of our outcome associated with a unit change in x_i, assuming no change in other inputs.

Applying an exponent as before, we have

$$\frac{P(y=1)}{P(y=0)} = e^{\beta_0 + \beta_1 x_1 + \beta_2 x_2 + \cdots + \beta_p x_p}$$
$$= e^{\beta_0} (e^{\beta_1})^{x_1} (e^{\beta_2})^{x_2} \dots (e^{\beta_p})^{x_p}$$

Therefore we can conclude that:

- e^{β_0} represents the odds of the outcome when all inputs are zero.
- Each e^{β_i} represents the *odds ratio* associated with a unit increase in x_i assuming no change in the other inputs (that is, a unit increase in x_i multiplies the odds of our outcome by e^{β_i}).

Let's put this into practice.

5.3.1 Running and interpreting a multivariate binomial logistic regression model

Let's use a binomial logistic regression model to understand how each of the three inputs in our `salespeople` data set influence the likelihood of promotion.

First, as we learned previously, it is good practice to convert the categorical `performance` variable to a dummy variable[4].

```
library(dummies)

# convert performance to dummy
perf_dummies <- dummies::dummy("performance", data = salespeople)


# replace in salespeople dataframe
salespeople_dummies <- cbind(
  salespeople[c("promoted", "sales", "customer_rate")],
  perf_dummies
)

# check it worked
head(salespeople_dummies)
```

---

[4]Note that most standard modeling functions have a built-in capability to deal with categorical variables, meaning that it's often not necessary to explicitly construct dummies. However, it is shown here for completion sake. You may wish to try running the subsequent code without explicitly constructing dummies, but note that constructing your own dummies gives you greater control over how they are labeled in any modeling output.

```
##   promoted sales customer_rate performance1 performance2 performance3 performance4
## 1        0   594          3.94            0            1            0            0
## 2        0   446          4.06            0            0            1            0
## 3        1   674          3.83            0            0            0            1
## 4        0   525          3.62            0            1            0            0
## 5        1   657          4.40            0            0            1            0
## 6        1   918          4.54            0            1            0            0
```

Now we can run our model (using the formula promoted ~ . to mean regressing promoted against everything else) and view our coefficients.

```
# run binomial glm
full_model <- glm(formula = "promoted ~ .",
                  family = "binomial",
                  data = salespeople_dummies)

# get coefficient summary
(coefs <- summary(full_model)$coefficients)
```

```
##                   Estimate   Std. Error     z value      Pr(>|z|)
## (Intercept)   -19.12443855 3.501115197  -5.46238483 4.697803e-08
## sales           0.04012425 0.006576429   6.10122119 1.052611e-09
## customer_rate  -1.11213130 0.482681585  -2.30406822 2.121881e-02
## performance1   -0.73449340 1.071963758  -0.68518492 4.932272e-01
## performance2   -0.47149387 0.933503552  -0.50507989 6.135027e-01
## performance3   -0.04953888 0.911614825  -0.05434189 9.566628e-01
```

Note how only three of the performance dummies have displayed. This is because everyone is in one of the four performance categories, so the model is using performance4 as the reference case. We can interpret each performance coefficient as the effect of a move to that performance category from performance4. We can already see from the last column of our coefficient summary—the coefficient p-values—that only sales and customer_rate meet the significance threshold of less than 0.05. Interestingly, it appears from the Estimate column that customer_rate has a negative effect on the log odds of promotion. For convenience, we can add an extra column to our coefficient summary to create the exponents of our estimated coefficients so that we can see the odds ratios. We can also remove columns that are less useful to us if we wish.

```
# create coefficient table with estimates, p-values and odds ratios
(full_coefs <- cbind(coefs[ ,c("Estimate", "Pr(>|z|)")],
                     odds_ratio = exp(full_model$coefficients)))
```

```
##                      Estimate      Pr(>|z|)      odds_ratio
## (Intercept)      -19.12443855  4.697803e-08  4.947227e-09
## sales              0.04012425  1.052611e-09  1.040940e+00
## customer_rate     -1.11213130  2.121881e-02  3.288573e-01
## performance1      -0.73449340  4.932272e-01  4.797484e-01
## performance2      -0.47149387  6.135027e-01  6.240693e-01
## performance3      -0.04953888  9.566628e-01  9.516682e-01
```

Now we can interpret our model as follows:

- All else being equal, sales have a significant positive effect on the likelihood of promotion, with each additional thousand dollars of sales increasing the odds of promotion by 4%
- All else being equal, customer ratings have a significant negative effect on the likelihood of promotion, with one full rating higher associated with 67% lower odds of promotion
- All else being equal, performance ratings have no significant effect on the likelihood of promotion

The second conclusion may appear counter-intuitive, but remember from our pairplot in Section 5.1.3 that there is already moderate correlation between sales and customer ratings, and this model will be controlling for that relationship. Recall that our odds ratios act *assuming all other variables are the same*. Therefore, if two individuals have the same sales and performance ratings, the one with the lower customer rating is more likely to have been promoted. Similarly, if two individuals have the same level of sales and the same customer rating, their performance rating will have no significant bearing on the likelihood of promotion.

Many analysts will feel uncomfortable with stating these conclusions with too much precision, and therefore exponent confidence intervals can be calculated to provide a range for the odds ratios.

```
exp(confint(full_model))
```

```
##                        2.5 %         97.5 %
## (Intercept)      1.505306e-12  1.750716e-06
## sales            1.029762e+00  1.057214e+00
## customer_rate    1.141645e-01  7.793018e-01
## performance1     5.345231e-02  3.824309e+00
## performance2     9.675452e-02  3.958066e+00
## performance3     1.591405e-01  5.976988e+00
## performance4              NA            NA
```

Therefore we can say that—all else being equal—every additional unit of sales increases the odds of promotion by between 3.0% and 5.7%, and every additional point in customer rating decreases the odds of promotion by between 22% and 89%.

Similar to other regression models, the unit scale needs to be taken into consideration during interpretation. On first sight, a decrease of up to 89% in odds seems a lot more important than an increase of up to 5.7% in odds. However, the increase of up to 5.7% is for one unit ($1000) in many thousands of sales units, and over 10 or 100 additional units can have a substantial compound effect on odds of promotion. The decrease of up to 89% is on a full customer rating point on a scale of only 4 full points.

5.3.2 Understanding the fit and goodness-of-fit of a binomial logistic regression model

Understanding the fit of a binomial logistic regression model is not straightforward and is sometimes controversial. Before we discuss this, let's simplify our model based on our learning that the performance data has no significant effect on the outcome.

```
# simplify model
simpler_model <- glm(formula = promoted ~ sales + customer_rate,
                     family = "binomial",
                     data = salespeople)
```

As in the previous chapter, again we have the luxury of a three-dimensional model, so we can visualize it in Figure 5.8, revealing a 3D sigmoid curve which 'twists' to reflect the relative influence of sales and customer_rate on the outcome.

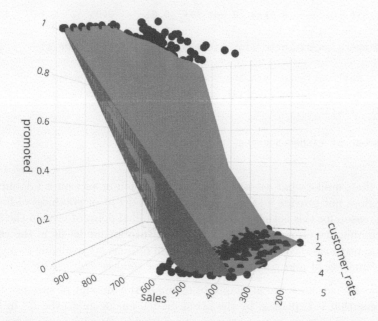

FIGURE 5.8: 3D visualization of the fitted `simpler_model` against the `sales-people` data

Now let's look at the summary of our `simpler_model`.

```
summary(simpler_model)
```

```
##
## Call:
## glm(formula = promoted ~ sales + customer_rate, family = "binomial",
##     data = salespeople)
##
## Deviance Residuals:
##      Min        1Q     Median        3Q        Max
## -2.02984  -0.09256  -0.02070   0.00874    3.06380
##
## Coefficients:
##                 Estimate Std. Error z value Pr(>|z|)
## (Intercept)   -19.517689   3.346762  -5.832 5.48e-09 ***
## sales           0.040389   0.006525   6.190 6.03e-10 ***
## customer_rate  -1.122064   0.466958  -2.403   0.0163 *
## ---
```

```
## Signif. codes:  0 '***' 0.001 '**' 0.01 '*' 0.05 '.' 0.1 ' ' 1
##
## (Dispersion parameter for binomial family taken to be 1)
##
##     Null deviance: 440.303  on 349  degrees of freedom
## Residual deviance:  65.131  on 347  degrees of freedom
## AIC: 71.131
##
## Number of Fisher Scoring iterations: 8
```

Note that, unlike what we saw for linear regression in Section 4.3.3, our summary does not provide a statistic on overall model fit or goodness-of-fit. The main reason for this is that there is no clear unified point of view in the statistics community on a single appropriate measure for model fit in the case of logistic regression. Nevertheless, a number of options are available to analysts for estimating fit and goodness-of-fit for these models.

Pseudo-R^2 measures are attempts to estimate the amount of variance in the outcome that is explained by the fitted model, analogous to the R^2 in linear regression. There are numerous variants of pseudo-R^2 with some of the most common listed here:

- McFadden's R^2 works by comparing the likelihood function of the fitted model with that of a random model and using this to estimate the explained variance in the outcome.
- Cox and Snell's R^2 works by applying a 'sum of squares' analogy to the likelihood functions to align more closely with the precise methodology for calculating R^2 in linear regression. However, this usually means that the maximum value is less than 1 and in certain circumstances substantially less than 1, which can be problematic and unintuitive for an R^2.
- Nagelkerke's R^2 resolves the issue with the upper bound for Cox and Snell by dividing Cox and Snell's R^2 by its upper bound. This restores an intuitive scale with a maximum of 1, but is considered somewhat arbitrary with limited theoretical foundation.
- Tjur's R^2 is a more recent and simpler concept. It is defined as simply the absolute difference between the predicted probabilities of the positive observations and those of the negative observations.

Standard modeling functions generally do not offer the calculation of pseudo-R^2 as standard, but numerous methods are available for their calculation. For example:

```
library(DescTools)
DescTools::PseudoR2(
  simpler_model,
  which = c("McFadden", "CoxSnell", "Nagelkerke", "Tjur")
)
```

```
##    McFadden   CoxSnell Nagelkerke       Tjur
## 0.8520759  0.6576490  0.9187858  0.8784834
```

We see that the Cox and Snell variant is notably lower than the other estimates, which is consistent with the known issues with its upper bound. However, the other estimates are reasonably aligned and suggest a strong fit.

Goodness-of-fit tests for logistic regression models compare the predictions to the observed outcome and test the null hypothesis that they are similar. This means that, unlike in linear regression, a low p-value indicates a poor fit. One commonly used method is the Hosmer-Lemeshow test, which divides the observations into a number of groups (usually 10) according to their fitted probabilities, calculates the proportion of each group that is positive and then compares this to the expected proportions based on the model prediction using a Chi-squared test. However, this method has limitations. It is particularly problematic for situations where there is a low sample size and can return highly varied results based on the number of groups used. It is therefore recommended to use a range of goodness-of-fit tests, and not rely entirely on any one specific approach.

In R, the LogisticDx package offers a range of diagnostic tools for logistic regression models, and is recommended for exploration. Here is an example using the gof() function for assessing goodness-of-fit.

```
library(LogisticDx)

# get range of goodness-of-fit diagnostics
simpler_model_diagnostics <- LogisticDx::gof(simpler_model,
                                              plotROC = FALSE)

# returns a list
names(simpler_model_diagnostics)
```

```
## [1] "ct"     "chiSq" "ctHL"   "gof"     "R2"     "auc"
```

The gof object in this list provides a range of variants of goodness-of-fit statistics.

```
# in our case we are interested in goodness-of-fit statistics
simpler_model_diagnostics$gof
```

```
##           test  stat        val df        pVal
## 1:          HL chiSq  3.44576058  8 0.903358158
## 2:         mHL     F  2.74709957  8 0.005971045
## 3:        OsRo     Z -0.02415249 NA 0.980730971
## 4: SstPgeq0.5     Z  0.88656856 NA 0.375311227
## 5:    SstPl0.5     Z  0.96819352 NA 0.332947728
## 6:     SstBoth chiSq  1.72340251  2 0.422442787
## 7: SllPgeq0.5 chiSq  1.85473814  1 0.173233325
## 8:    SllPl0.5 chiSq  0.68570870  1 0.407627859
## 9:     SllBoth chiSq  1.86640617  2 0.393291943
```

This confirms that almost all tests, including the Hosmer-Lemeshow test, which is the first in the list, suggest a fit for our model.

Various measures of predictive accuracy can also be used to assess a binomial logistic regression model in a predictive context, such as precision, recall and ROC-curve analysis. These are particularly suited for implementations of logistic regression models as predictive classifiers in a Machine Learning context, a topic which is outside the scope of this book. However, a recommended source for a deeper treatment of goodness-of-fit tests for logistic regression models is Hosmer, Lemeshow, and Sturdivant (2013).

5.3.3   Model parsimony

We saw that in both our linear regression and our logistic regression approach, we decided to drop variables from our model when we determined that they had no significant effect on the outcome. The principle of *Occam's Razor* states that—all else being equal—the simplest explanation is the best. In this sense, a model that contains information that does not contribute to its primary inference objective is more complex than it needs to be. Such a model increases the communication burden in explaining its results to others, with no notable analytic benefit in return.

Parsimony describes the concept of being careful with resources or with information. A model could be described as more parsimonious if it can achieve the same (or very close to the same) fit with a smaller number of inputs. The *Akaike Information Criterion* or *AIC* is a measure of model parsimony that is computed for log-likelihood models like logistic regression models, with a lower AIC indicating a more parsimonious model. AIC is often calculated as standard in summary reports of logistic regression models but can also be calculated independently. Let's compare the different iterations of our model in this chapter using AIC.

```
# sales only model
AIC(sales_model)
```

```
## [1] 76.49508
```

```
# sales and customer rating model
AIC(simpler_model)
```

```
## [1] 71.13145
```

```
# model with all inputs
AIC(full_model)
```

```
## [1] 76.37433
```

We can see that the model which is limited to our two significant inputs—sales and customer rating—is determined to be the most parsimonious model according to the AIC. Note that the AIC should not be used to interpret model quality or confidence—it is possible that the lowest AIC might still be a very poor fit.

Model parsimony becomes a substantial concern when there is a large number of input variables. As a general rule, the more input variables there are in a model the greater the chance that the model will be difficult to interpret clearly, and the greater the risk of measurement problems, such as multicollinearity. Analysts who are eager to please their customers, clients, professors or bosses can easily be tempted to think up new potential inputs to their model, often derived mathematically from measures that are already inputs in the model. Before long the model is too complex, and in extreme cases there are more inputs than there are observations. The primary way to manage model complexity is to exercise caution in selecting model inputs. When large numbers of inputs are unavoidable, coefficient regularization methods such as LASSO regression can help with model parsimony.

5.4   Other considerations in binomial logistic regression

To predict from new data, just use the predict() function as in the previous
chapter. This function recognizes the type of model being used—in this case
a generalized linear model—and adjusts its prediction approach accordingly.
In particular, if you want to return the probability of the new observations
being promoted, you need to use type = "response" as an argument.

```
# define new observations
(new_data <- data.frame(sales = c(420, 510, 710),
                        customer_rate = c(3.4, 2.3, 4.2)))
```

```
##    sales customer_rate
## 1    420           3.4
## 2    510           2.3
## 3    710           4.2
```

```
# predict probability of promotion
predict(simpler_model, new_data, type = "response")
```

```
##          1          2          3
## 0.00171007 0.18238565 0.98840506
```

Many of the principles covered in the previous chapter on linear regression are
equally important in logistic regression. For example, input variables should
be managed in a similar way. Collinearity and multicollinearity should be of
concern. Interaction of input variables can be modeled. For the most part,
analysts should be aware of the fundamental mathematical transformations
which take place in a logistic regression model when they consider some of
these issues (another reason to ensure that the mathematics covered earlier
in this chapter is well understood). For example, while coefficients in linear
regression have a direct additive impact on y, in logistic regression they have
a direct additive impact on the log odds of y, or alternatively their exponents
have a direct multiplicative impact on the odds of y. Therefore coefficient
overestimation such as that which can occur when collinearity is not managed
can result in inferences that could substantially overstate the importance or
effect of input variables.

Because of the binary nature of our outcome variable, the residuals of a logistic regression model have limited direct application to the problem being studied. In practical contexts the residuals of logistic regression models are rarely examined, but they can be useful in identifying outliers or particularly influential observations and in assessing goodness-of-fit. When residuals are examined, they need to be transformed in order to be analyzed appropriately. For example, the *Pearson residual* is a standardized form of residual from logistic regression which can be expected to have a normal distribution over large-enough samples. We can see in Figure 5.9 that this is the case for our `simpler_model`, but that there are a small number of substantial underestimates in our model. A good source of further learning on diagnostics of logistic regression models is Menard (2010).

```
d <- density(residuals(simpler_model, "pearson"))
plot(d, main= "")
```

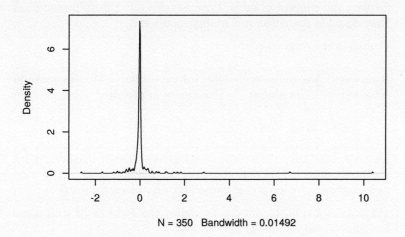

N = 350   Bandwidth = 0.01492

FIGURE 5.9: Distribution of Pearson residuals in `simpler_model`

5.5   Learning exercises

5.5.1   Discussion questions

1. Draw the shape of a logistic function. Describe the three population growth phases it was originally intended to model.

2. Explain why the logistic function is useful to statisticians in modeling.

3. In the formula for the logistic function in Section 5.1.1, what might be a common value for L in probabilistic applications? Why?

4. What types of problems are suitable for logistic regression modeling?

5. Can you think of some modeling scenarios in your work or studies that could use a logistic regression approach?

6. Explain the concept of odds. How do odds differ from probability? How do odds change as probability increases?

7. Complete the following:

a. If an event has a 1% probability of occurring, a 10% increase in odds results in an almost ___% increase in probability.
b. If an event has a 99% probability of occurring, a 10% increase in odds results in an almost ___% increase in probability.

8. Describe how the coefficients of a logistic regression model affect the fitted outcome. If β is a coefficient estimate, how is the odds ratio associated with β calculated and what does it mean?

9. What are some of the options for determining the fit of a binomial logistic regression model?

10. Describe the concept of model parsimony. What measure is commonly used to determine the most parsimonious logistic regression model?

5.5.2   Data exercises

A nature preservation charity has asked you to analyze some data to help them understand the features of those members of the public who donated in a given month. Load the charity_donation data set via the peopleanalyticsdata package or download it from the internet[5]. It contains the following data:

---

[5]http://peopleanalytics-regression-book.org/data/charity_donation.csv

- n_donations: The total number of times the individual donated previous to the month being studied.
- total_donations: The total amount of money donated by the individual previous to the month being studied
- time_donating: The number of months between the first donation and the month being studied
- recent_donation: Whether or not the individual donated in the month being studied
- last_donation: The number of months between the most recent previous donation and the month being studied
- gender: The gender of the individual
- reside: Whether the person resides in an Urban or Rural Domestic location or Overseas
- age: The age of the individual

1. View the data and obtain statistical summaries. Ensure data types are appropriate and there is no missing data. Determine the outcome and input variables.
2. Using a pairplot or by plotting or correlating selected fields, try to hypothesize which variables may be significant in explaining who recently donated.
3. Run a binomial logistic regression model using all input fields. Determine which input variables have a significant effect on the outcome and the direction of that effect.
4. Calculate the odds ratios for the significant variables and explain their impact on the outcome.
5. Check for collinearity or multicollinearity in your model using methods from previous chapters.
6. Experiment with model parsimony by reducing input variables that do not have a significant impact on the outcome. Decide on the most parsimonious model.
7. Calculate a variety of Pseudo-R^2 variants for your model. How would you explain these to someone with no statistics expertise?
8. Report the conclusions of your modeling exercise to the charity by writing a simple explanation that assumes no knowledge of statistics.
9. **Extension:** Using a variety of methods of your choice, test the hypothesis that your model fits the data. How conclusive are your tests?

6

Multinomial Logistic Regression for Nominal Category Outcomes

In the previous chapter we looked at how to model a binary or dichotomous outcome using a logistic function. In this chapter we look at how to extend this to the case when the outcome has a number of categories that do not have any order to them. When an outcome has this nominal categorical form, it does not have a sense of direction. There is no 'better' or 'worse', no 'higher' or 'lower', there is only 'different'.

6.1 When to use it

6.1.1 Intuition for multinomial logistic regression

A binary or dichotomous outcome like we studied in the previous chapter is already in fact a nominal outcome with two categories, so in principle we already have the basic technology with which to study this problem. That said, the way we approach the problem can differ according to the types of inferences we wish to make.

If we only wish to make inferences about the choice of each specific category— what drives whether an observation is in Category A versus the others, or Category B versus the others—then we have the option of running separate binomial logistic regression models on a 'one versus the rest' basis. In this case we can refine our model differently for each category, eliminating variables that are not significant in determining membership of that category. This could potentially lead to models being defined differently for different target outcome categories. Notably, there will be no common comparison category between these models. This is sometimes called a *stratified* approach.

However, in many studies there is a need for a 'reference' category to better understand the relative odds of category membership. For example, in clinical settings the relative risk factors for different clinical outcomes can only be understood relative to a reference (usually that of the 'most healthy' or 'most

DOI: 10.1201/9781003194156-6

recovered' patients)[1]. In organizational settings, one can imagine that the odds of different types of mid-tenure career path changes could only be well understood relative to a reference career path (probably the most common one). While this approach would still be founded on binomial models, the reference points of these models are different; we would need to make decisions on refining the model differently, and we interpret the coefficients in a different way.

In this chapter we will briefly look at the stratified approach (which is effectively a repetition of work done in the previous chapter) before focusing more intently on how we construct models and make inferences using a multinomial approach.

6.1.2   Use cases for multinomial logistic regression

Multinomial logistic regression is appropriate for any situation where a limited number of outcome categories (more than two) are being modeled and where those outcome categories have no order. An underlying assumption is the independence of irrelevant alternatives (IIA). Otherwise stated, this assumption means that there is no other alternative for the outcome that, if included, would disproportionately influence the membership of one of the other categories[2]. In cases where this assumption is violated, one could choose to take a stratified approach, or attempt hierarchical or nested multinomial model alternatives, which are beyond the scope of this book.

Examples of typical situations that might be modeled by multinomial logistic regression include:

1. Modeling voting choice in elections with multiple candidates
2. Modeling choice of career options by students
3. Modeling choice of benefit options by employees

6.1.3   Walkthrough example

You are an analyst at a large technology company. The company recently introduced a new health insurance provider for its employees. At the beginning of the year the employees had to choose one of three different health plan

---

[1] In Hosmer, Lemeshow, and Sturdivant (2013), a good example is provided where the outcome is the placement of psychiatric patients in various forms of aftercare, with Outpatient Care as the reference.

[2] Put differently, it assumes that adding or removing any other available alternative would affect the odds of the other alternatives in equal proportion. It has been shown that there have been many studies that proceeded with a multinomial approach despite the violation of this assumption.

products from this provider to best suit their needs. You have been asked to determine which factors influenced the choice in product.

The `health_insurance` data set consists of the following fields:

- `product`: The choice of product of the individual—A, B or C
- `age`: The age of the individual when they made the choice
- `gender`: The gender of the individual as stated when they made the choice
- `household`: The number of people living with the individual in the same household at the time of the choice
- `position_level`: Position level in the company at the time they made the choice, where 1 is is the lowest and 5 is the highest
- `absent`: The number of days the individual was absent from work in the year prior to the choice

First we load the data and take a look at it briefly.

```
# if needed, download health_insurance data
url <- "http://peopleanalytics-regression-book.org/data/health_insurance.csv"
health_insurance <- read.csv(url)
```

```
# view first few rows
head(health_insurance)
```

```
##   product age household position_level gender absent
## 1       C  57         2              2   Male     10
## 2       A  21         7              2   Male      7
## 3       C  66         7              2   Male      1
## 4       A  36         4              2 Female      6
## 5       A  23         0              2   Male     11
## 6       A  31         5              1   Male     14
```

```
# view structure
str(health_insurance)
```

```
## 'data.frame':   1453 obs. of  6 variables:
##  $ product       : chr  "C" "A" "C" "A" ...
##  $ age           : int  57 21 66 36 23 31 37 37 55 66 ...
##  $ household     : int  2 7 7 4 0 5 3 0 3 2 ...
##  $ position_level: int  2 2 2 2 2 1 3 3 3 4 ...
##  $ gender        : chr  "Male" "Male" "Male" "Female" ...
##  $ absent        : int  10 7 1 6 11 14 12 25 3 18 ...
```

It looks like two of these columns should be converted to factor—product and
gender—so let's do that and then run a pairplot for a quick overview of any
patterns, which can be seen in Figure 6.1.

```
library(GGally)

# convert product and gender to factors
health_insurance$product <- as.factor(health_insurance$product)
health_insurance$gender <- as.factor(health_insurance$gender)

GGally::ggpairs(health_insurance)
```

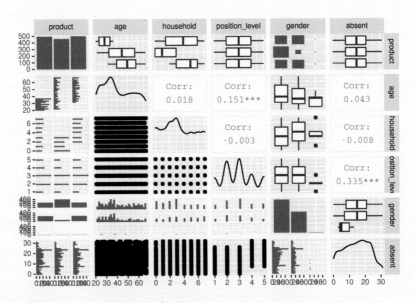

FIGURE 6.1: Pairplot of the `health_insurance` data set

The data appears somewhat chaotic here. However, there are a few things
to note. Firstly, we notice that there is a relatively even spread in choice
between the products. We also notice that age seems to be playing a role

in product choice. There are also some mild-to-moderate correlations in the data—in particular between age and position_level, and between absent and position_level. However, this problem is clearly more complex than we can determine from a bivariate perspective.

6.2 Running stratified binomial models

One approach to this problem is to look at each product choice and treat it as an independent binomial logistic regression model, modeling that choice against an alternative of all other choices. Each such model may help us describe the dynamics of the choice of a specific product, but we have to be careful in making conclusions about the overall choice between the three products. Running stratified models would not be very efficient if we had a wider range of choices for our outcome, but since we only have three possible choices here, it is reasonable to take this route.

6.2.1 Modeling the choice of Product A versus other products

Let's first create and refine a binomial model for the choice of Product A.

```
library(dummies)

# create dummies for product choice outcome
dummy_product <- dummies::dummy("product", data = health_insurance)

# combine to original set
health_insurance <- cbind(health_insurance, dummy_product)

# run a binomial model for the Product A dummy against
# all input variables (let glm() handle dummy input variables)
A_model <- glm(
  formula = productA ~ age + gender + household +
    position_level + absent,
  data = health_insurance,
  family = "binomial"
)
```

```
# summary
summary(A_model)
```

```
##
## Call:
## glm(formula = productA ~ age + gender + household + position_level +
##     absent, family = "binomial", data = health_insurance)
##
## Deviance Residuals:
##     Min       1Q    Median        3Q       Max
## -2.19640  -0.43691  -0.07051   0.46304   2.37416
##
## Coefficients:
##                   Estimate Std. Error z value Pr(>|z|)
## (Intercept)       5.873634   0.453041  12.965  < 2e-16 ***
## age              -0.239814   0.013945 -17.197  < 2e-16 ***
## genderMale        0.845978   0.168237   5.028 4.94e-07 ***
## genderNon-binary  0.222521   1.246591   0.179    0.858
## household         0.240205   0.037358   6.430 1.28e-10 ***
## position_level    0.321497   0.071770   4.480 7.48e-06 ***
## absent           -0.003751   0.010753  -0.349    0.727
## ---
## Signif. codes:  0 '***' 0.001 '**' 0.01 '*' 0.05 '.' 0.1 ' ' 1
##
## (Dispersion parameter for binomial family taken to be 1)
##
##     Null deviance: 1864.15  on 1452  degrees of freedom
## Residual deviance:  940.92  on 1446  degrees of freedom
## AIC: 954.92
##
## Number of Fisher Scoring iterations: 6
```

We see that all variables except absent seem to play a significant role in the choice of Product A. All else being equal, being older makes the choice of Product A less likely. Males are more likely to choose Product A, and larger households and higher position levels also make the choice of Product A more likely. Based on this, we can consider simplifying our model to remove absent. We can also calculate odds ratios and perform some model diagnostics if we wish, similar to how we approached the problem in the previous chapter.

These results need to be interpreted carefully. For example, the odds ratios for the Product A choice based on a simplified model are as follows:

```
# simpler model
A_simple <- glm(
  formula = productA ~ age + household + gender + position_level,
  data = health_insurance
)
```

```
# view odds ratio as a data frame
as.data.frame(exp(A_simple$coefficients))
```

```
##                    exp(A_simple$coefficients)
## (Intercept)                         2.8212501
## age                                 0.9776023
## household                           1.0253707
## genderMale                          1.0973280
## genderNon-binary                    0.9697639
## position_level                      1.0397906
```

As an example, and as a reminder from our previous chapter, we interpret the odds ratio for age as follows: all else being equal, every additional year of age is associated with an approximately 2.2% decrease in the odds of choosing Product A over the other products.

6.2.2 Modeling other choices

In a similar way we can produce two other models, representing the choice of Products B and C. These models produce similar significant variables, except that position_level does not appear to be significant in the choice of Product C. If we simplify all our three models we will have a slightly differently defined model for the choice of Product C versus our models for the other two product choices. However, we can conclude in general that the only input variable that seems to be non-significant across all choices of product is absent.

6.3 Running a multinomial regression model

An alternative to running separate binary stratified models is to run a multinomial logistic regression model. A multinomial logistic model will base itself from a defined reference category, and run a generalized linear model on the log-odds of membership of each of the other categories versus the reference category. Due to its extensive use in epidemiology and medicine, this is often known as the *relative risk* of one category compared to the reference category. Mathematically speaking, if X is the vector of input variables, and y takes the value A, B or C, with A as the reference, a multinomial logistic regression model will calculate:

$$\ln \left(\frac{P(y = B)}{P(y = A)} \right) = \alpha X$$

and

$$\ln \left(\frac{P(y = C)}{P(y = A)} \right) = \beta X$$

for different vectors of coefficients α and β.

6.3.1 Defining a reference level and running the model

The nnet package in R contains a multinom() function for running a multinomial logistic regression model using neural network technology[3]. Before we can run the model we need to make sure our reference level is defined.

```
# define reference by ensuring it is the first level of the factor
health_insurance$product <- relevel(health_insurance$product, ref = "A")

# check that A is now our reference
levels(health_insurance$product)
```

```
## [1] "A" "B" "C"
```

Once the reference outcome is defined, the multinom() function from the nnet package will run a series of binomial models comparing the reference to each of the other categories.

First we will calculate our multinomial logistic regression model.

---

[3] Neural networks are computational structures which consist of a network of nodes, each of which take an input and perform a mathematical function to return an output onward in the network. Most commonly they are used in deep learning, but a simple neural network here can model these different categories using a logistic function.

```
library(nnet)

multi_model <- multinom(
  formula = product ~ age + gender + household +
    position_level + absent,
  data = health_insurance
)
```

Now we will look at a summary of the results.

```
summary(multi_model)
```

```
## Call:
## multinom(formula = product ~ age + gender + household + position_level +
##     absent, data = health_insurance)
##
## Coefficients:
##    (Intercept)       age genderMale genderNon-binary household position_level       absent
## B    -4.60100 0.2436645 -2.38259765        0.2523409 -0.9677237     -0.4153040 0.011676034
## C   -10.22617 0.2698141  0.09670752       -1.2715643  0.2043568     -0.2135843 0.003263631
##
## Std. Errors:
##    (Intercept)        age genderMale genderNon-binary household position_level     absent
## B    0.5105532 0.01543139  0.2324262         1.226141 0.06943089     0.08916739 0.01298141
## C    0.6197408 0.01567034  0.1954353         2.036273 0.04960655     0.08226087 0.01241814
##
## Residual Deviance: 1489.365
## AIC: 1517.365
```

Notice that the output of `summary(multi_model)` is much less detailed than for our standard binomial models, and it effectively just delivers the coefficients and standard errors of the two models against the reference. To determine whether specific input variables are significant we will need to calculate the p-values of the coefficients manually by calculating the z-statistics and converting (we covered this hypothesis testing methodology in Section 3.3.1).

```
# calculate z-statistics of coefficients
z_stats <- summary(multi_model)$coefficients/
  summary(multi_model)$standard.errors

# convert to p-values
p_values <- (1 - pnorm(abs(z_stats)))*2


# display p-values in transposed data frame
data.frame(t(p_values))
```

```
##                            B             C
## (Intercept)     0.000000e+00  0.000000e+00
## age             0.000000e+00  0.000000e+00
## genderMale      0.000000e+00  6.207192e-01
## genderNon-binary 8.369465e-01 5.323278e-01
## household       0.000000e+00  3.796088e-05
## position_level  3.199529e-06  9.419906e-03
## absent          3.684170e-01  7.926958e-01
```

6.3.2   Interpreting the model

This confirms that all variables except absent play a role in the choice between all products relative to a reference of Product A. We can also calculate odds ratios as before.

```
# display odds ratios in transposed data frame
odds_ratios <- exp(summary(multi_model)$coefficients)
data.frame(t(odds_ratios))
```

```
##                            B             C
## (Intercept)       0.01004179  3.621021e-05
## age               1.27591615  1.309721e+00
## genderMale        0.09231048  1.101538e+00
## genderNon-binary  1.28703467  2.803927e-01
## household         0.37994694  1.226736e+00
## position_level    0.66013957  8.076841e-01
## absent            1.01174446  1.003269e+00
```

Here are some examples of how these odds ratios can be interpreted in the multinomial context (used in combination with the p-values above):

- All else being equal, every additional year of age increases the relative odds of selecting Product B versus Product A by approximately 28%, and increases the relative odds of selecting Product C versus Product A by approximately 31%
- All else being equal, being Male reduces the relative odds of selecting Product B relative to Product A by 91%.
- All else being equal, each additional household member deceases the odds of selecting Product B relative to Product A by 62%, and increases the odds of selecting Product C relative to Product A by 23%.

6.3.3 Changing the reference

It may be the case that someone would like to hear the odds ratios stated against the reference of an individual choosing Product B. For example, what are the odds ratios of Product C relative to a reference of Product B? One way to do this would be to change the reference and run the model again. Another option is to note that:

$$\frac{P(y = C)}{P(y = B)} = \frac{\frac{P(y=C)}{P(y=A)}}{\frac{P(y=B)}{P(y=A)}} = \frac{e^{\beta X}}{e^{\alpha X}} = e^{(\beta - \alpha)X}$$

Therefore

$$\ln\left(\frac{P(y = C)}{P(y = B)}\right) = (\beta - \alpha)X$$

This means we can obtain the coefficients of C against the reference of B by simply calculating the difference between the coefficients of C and B against the common reference of A. Let's do this.

```
# calculate difference between coefficients and view as column
coefs_c_to_b <- summary(multi_model)$coefficients[2, ] -
    summary(multi_model)$coefficients[1, ]

data.frame(coefs_c_to_b)
```

```
##                      coefs_c_to_b
## (Intercept)          -5.625169520
## age                   0.026149597
## genderMale            2.479305168
## genderNon-binary     -1.523905192
## household             1.172080452
## position_level        0.201719688
## absent               -0.008412403
```

If the number of categories in the outcome variable is limited, this can be an efficient way to obtain the model coefficients against various reference points without having to rerun models. However, to determine standard errors and p-values for these coefficients the model will need to be recalculated against the new reference.

6.4 Model simplification, fit and goodness-of-fit for multinomial logistic regression models

Simplifying a multinomial regression model needs to be done with care. In a binomial model, there is one set of coefficients and their p-values can be a strong guide to which variables can be removed safely. However, in multinomial models there are several sets of coefficients to consider.

6.4.1 Gradual safe elimination of variables

In Hosmer, Lemeshow, and Sturdivant (2013), a gradual process of elimination of variables is recommended to ensure that significant variables that confound each other in the different logistic models are not accidentally dropped from the final model. The recommended approach is as follows:

- Start with the variable with the least significant p-values in all sets of coefficients—in our case absent would be the obvious first candidate.
- Run the multinomial model without this variable.
- Test that none of the previous coefficients change by more than 20–25%.
- If there was no such change, safely remove the variable and proceed to the next non-significant variable.
- If there is such a change, retain the variable and proceed to the next non-significant variable.
- Stop when all non-significant variables have been tested.

In our case, we can compare the coefficients of the model with and without absent included and verify that the changes in the coefficients are not substantial.

```
# remove absent
simpler_multi_model <- multinom(
  formula = product ~ age + gender + household + position_level,
  data = health_insurance,
  model = TRUE
)
```

```
# view coefficients with absent
data.frame(t(summary(multi_model)$coefficients))
```

```
##                                B             C
## (Intercept)         -4.60099991 -10.226169428
## age                  0.24366447   0.269814063
## genderMale          -2.38259765   0.096707521
## genderNon-binary     0.25234087  -1.271564323
## household           -0.96772368   0.204356774
## position_level      -0.41530400  -0.213584308
## absent               0.01167603   0.003263631
```

```
# view coefficients without absent
data.frame(t(summary(simpler_multi_model)$coefficients))
```

```
##                             B            C
## (Intercept)        -4.5008999 -10.19269011
## age                 0.2433855   0.26976294
## genderMale         -2.3771342   0.09801281
## genderNon-binary    0.1712091  -1.29636779
## household          -0.9641956   0.20510806
## position_level     -0.3912014  -0.20908835
```

We can see that only genderNon-binary changed substantially, but we note that this is on an extremely small sample size and so will not have any effect on our model[4]. It therefore appears safe to remove absent. Furthermore, the Akaike Information Criterion is equally valid in multinomial models for evaluating model parsimony. Here we can calculate that the AIC of our model with and without absent is 1517.36 and 1514.25, respectively, confirming that the model without absent is marginally more parsimonious.

6.4.2 Model fit and goodness-of-fit

As with the binomial case, a variety of Pseudo-R^2 methods are available to assess the fit of a multinomial logistic regression model, although some of our previous variants (particularly Tjur) are not defined on models with more than two outcome categories.

---

[4]Removing insignificant dummy variables, or combining them to make simpler dummy variables can also be done. In the case of these observations of genderNon-binary, given the relatively small number of these observations in the data set, it does not harm the model to leave this variable included, safe in the knowledge that it has a minuscule effect

```
DescTools::PseudoR2(simpler_multi_model,
                    which = c("McFadden", "CoxSnell", "Nagelkerke"))
```

```
##    McFadden  CoxSnell Nagelkerke
## 0.5329175 0.6896945  0.7760413
```

Due to the fact that multinomial models have more than one set of coefficients, assessing goodness-of-fit is more challenging, and is still an area of intense research. The most approachable method to assess model confidence is the Hosmer-Lemeshow test mentioned in the previous chapter, which was extended in Fagerland, Hosmer, and Bofin (2008) for multinomial models. An implementation is available in the generalhoslem package in R. However, this version of the Hosmer-Lemeshow test is problematic for models with a small number of input variables (fewer than ten), and therefore we will not experiment with it here. For further exploration of this topic, Chapter 8 of Hosmer, Lemeshow, and Sturdivant (2013) is recommended, and for a more thorough treatment of the entire topic of categorical analytics, Agresti (2007) is an excellent companion.

6.5 Learning exercises

6.5.1 Discussion questions

1. Describe the difference between a stratified versus a multinomial approach to modeling an outcome with more than two nominal categories.
2. Describe how you would interpret the odds ratio of an input variable for a given category in a stratified modeling approach.
3. Describe what is meant by the 'reference' of a multinomial logistic regression model with at least three nominal outcome categories.
4. Describe how you would interpret the odds ratio of an input variable for a given category in a multinomial modeling approach.
5. Given a multinomial logistic regression model with outcome categories A, B, C and D and reference category A, describe two ways to determine the coefficients of a multinomial logistic regression model with reference category C.
6. Describe a process for safely simplifying a multinomial logistic regression model by removing input variables.

6.5.2 Data exercises

Use the same `health_insurance` data set from this chapter to answer these questions.

1. Complete the full stratified approach to modeling the three product choices that was started in Section 6.2. Calculate the coefficients, odds ratios and p-values in each case.
2. Carefully write down your interpretation of the odds ratios from the previous question.
3. Run a multinomial logistic regression model on the `product` outcome using Product B as reference. Calculate the coefficients, ratios and p-values in each case.
4. Verify that the coefficients for Product C against reference Product B matches those calculated in Section 6.3.3.
5. Carefully write down your interpretation of the odds ratios calculated in the previous question.
6. Use the process described in Section 6.4.1 to simplify the multinomial model in Question 3.

7

Proportional Odds Logistic Regression for Ordered Category Outcomes

Often our outcomes will be categorical in nature, but they will also have an order to them. These are sometimes known as *ordinal* outcomes. Some very common examples of this include ratings of some form, such as job performance ratings or survey responses on Likert scales. The appropriate modeling approach for these outcome types is ordinal logistic regression. Surprisingly, this approach is frequently not understood or adopted by analysts. Often they treat the outcome as a continuous variable and perform simple linear regression, which can lead to wildly inaccurate inferences. Given the prevalence of ordinal outcomes in people analytics, it would serve analysts well to know how to run ordinal logistic regression models, how to interpret them and how to confirm their validity.

In fact, there are numerous known ways to approach the inferential modeling of ordinal outcomes, all of which build on the theory of linear, binomial and multinomial regression which we covered in previous chapters. In this chapter, we will focus on the most commonly adopted approach: *proportional odds* logistic regression. Proportional odds models (sometimes known as constrained cumulative logistic models) are more attractive than other approaches because of their ease of interpretation but cannot be used blindly without important checking of underlying assumptions.

7.1 When to use it

7.1.1 Intuition for proportional odds logistic regression

Ordinal outcomes can be considered to be suitable for an approach somewhere 'between' linear regression and multinomial regression. In common with linear regression, we can consider our outcome to increase or decrease dependent on our inputs. However, unlike linear regression the increase and decrease is 'stepwise' rather than continuous, and we do not know that the difference between the steps is the same across the scale. In medical settings, the difference between moving from a healthy to an early-stage disease may not be equivalent to moving from an early-stage disease to an intermediate- or advanced-stage. Equally, it may be a much bigger psychological step for an individual to say that they are very dissatisfied in their work than it is to say that they are very satisfied in their work. In this sense, we are analyzing categorical outcomes similar to a multinomial approach.

To formalize this intuition, we can imagine a latent version of our outcome variable that takes a continuous form, and where the categories are formed at specific cutoff points on that continuous variable. For example, if our outcome variable y represents survey responses on an ordinal Likert scale of 1 to 5, we can imagine we are actually dealing with a continuous variable y' along with four increasing 'cutoff points' for y' at τ_1, τ_2, τ_3 and τ_4. We then define each ordinal category as follows: $y = 1$ corresponds to $y' \leq \tau_1$, $y \leq 2$ to $y' \leq \tau_2$, $y \leq 3$ to $y' \leq \tau_3$ and $y \leq 4$ to $y' \leq \tau_4$. Further, at each such cutoff τ_k, we assume that the probability $P(y > \tau_k)$ takes the form of a logistic function. Therefore, in the proportional odds model, we 'divide' the probability space at each level of the outcome variable and consider each as a binomial logistic regression model. For example, at rating 3, we generate a binomial logistic regression model of $P(y > \tau_3)$, as illustrated in Figure 7.1.

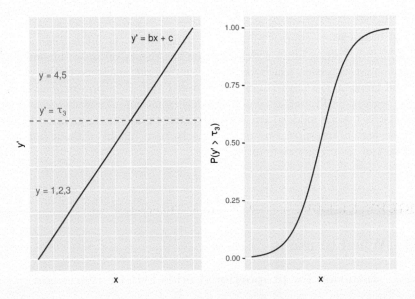

FIGURE 7.1: Proportional odds model illustration for a 5-point Likert survey scale outcome greater than 3 on a single input variable. Each cutoff point τ_k in the latent continuous outcome variable y' gives rise to a binomial logistic function.

This approach leads to a highly interpretable model that provides a single set of coefficients that are agnostic to the outcome category. For example, we can say that each unit increase in input variable x increases the odds of y being in a *higher category* by a certain ratio.

7.1.2 Use cases for proportional odds logistic regression

Proportional odds logistic regression can be used when there are more than two outcome categories that have an order. An important underlying assumption is that no input variable has a disproportionate effect on a specific level of the outcome variable. This is known as the proportional odds assumption. Referring to Figure 7.1, this assumption means that the 'slope' of the logistic function is the same for all category cutoffs[1]. If this assumption is violated, we cannot reduce the coefficients of the model to a single set across all outcome categories, and this modeling approach fails. Therefore, testing the proportional odds assumption is an important validation step for anyone running this type of model.

Examples of problems that can utilize a proportional odds logistic regression approach include:

1. Understanding the factors associated with higher ratings in an employee survey on a Likert scale
2. Understanding the factors associated with higher job performance ratings on an ordinal performance scale
3. Understanding the factors associated with voting preference in a ranked preference voting system (for example, proportional representation systems)

7.1.3 Walkthrough example

You are an analyst for a sports broadcaster who is doing a feature on player discipline in professional soccer games. To prepare for the feature, you have been asked to verify whether certain metrics are significant in influencing the extent to which a player will be disciplined by the referee for unfair or dangerous play in a game. You have been provided with data on over 2000 different players in different games, and the data contains these fields:

- `discipline`: A record of the maximum discipline taken by the referee against the player in the game. 'None' means no discipline was taken, 'Yellow' means the player was issued a yellow card (warned), 'Red' means the player was issued a red card and ordered off the field of play.

---

[1]This also leads to another term for the assumption—the *parallel regression* assumption.

- `n_yellow_25` is the total number of yellow cards issued to the player in the previous 25 games they played prior to this game.
- `n_red_25` is the total number of red cards issued to the player in the previous 25 games they played prior to this game.
- `position` is the playing position of the player in the game: 'D' is defense (including goalkeeper), 'M' is midfield and 'S' is striker/attacker.
- `level` is the skill level of the competition in which the game took place, with 1 being higher and 2 being lower.
- `country` is the country in which the game took place—England or Germany.
- `result` is the result of the game for the team of the player—'W' is win, 'L' is lose, 'D' is a draw/tie.

Let's download the `soccer` data set and take a quick look at it.

```
# if needed, download data
url <- "http://peopleanalytics-regression-book.org/data/soccer.csv"
soccer <- read.csv(url)
```

```
head(soccer)
```

```
##    discipline n_yellow_25 n_red_25 position result country level
## 1        None           4        1        S      D England     1
## 2        None           2        2        D      W England     2
## 3        None           2        1        M      D England     1
## 4        None           2        1        M      L Germany     1
## 5        None           2        0        S      W Germany     1
## 6        None           3        2        M      W England     1
```

Let's also take a look at the structure of the data.

```
str(soccer)
```

```
## 'data.frame':    2291 obs. of  7 variables:
##  $ discipline : chr  "None" "None" "None" "None" ...
##  $ n_yellow_25: int  4 2 2 2 2 3 4 3 4 3 ...
##  $ n_red_25   : int  1 2 1 1 0 2 2 0 3 3 ...
##  $ position   : chr  "S" "D" "M" "M" ...
##  $ result     : chr  "D" "W" "D" "L" ...
##  $ country    : chr  "England" "England" "England" "Germany" ...
##  $ level      : int  1 2 1 1 1 1 2 1 1 1 ...
```

We see that there are numerous fields that need to be converted to factors before we can model them. Firstly, our outcome of interest is discipline and this needs to be an ordered factor, which we can choose to increase with the seriousness of the disciplinary action.

```r
# convert discipline to ordered factor
soccer$discipline <- ordered(soccer$discipline,
                       levels = c("None", "Yellow", "Red"))

# check conversion
str(soccer)
```

```
## 'data.frame':    2291 obs. of  7 variables:
##  $ discipline : Ord.factor w/ 3 levels "None"<"Yellow"<..: 1 1 1 1 1 1 1 1 1 1 ...
##  $ n_yellow_25: int  4 2 2 2 2 3 4 3 4 3 ...
##  $ n_red_25   : int  1 2 1 1 0 2 2 0 3 3 ...
##  $ position   : chr  "S" "D" "M" "M" ...
##  $ result     : chr  "D" "W" "D" "L" ...
##  $ country    : chr  "England" "England" "England" "Germany" ...
##  $ level      : int  1 2 1 1 1 1 2 1 1 1 ...
```

We also know that position, country, result and level are categorical, so we convert them to factors. We could in fact choose to convert result and level into ordered factors if we so wish, but this is not necessary for input variables, and the results are usually a little bit easier to read as nominal factors.

```r
# apply as.factor to four columns
cats <- c("position", "country", "result", "level")
soccer[ ,cats] <- lapply(soccer[ ,cats], as.factor)

# check again
str(soccer)
```

```
## 'data.frame':    2291 obs. of  7 variables:
## $ discipline : Ord.factor w/ 3 levels "None"<"Yellow"<..: 1 1 1 1 1 1 1 1 1 1 ...
## $ n_yellow_25: int  4 2 2 2 2 3 4 3 4 3 ...
## $ n_red_25   : int  1 2 1 1 0 2 2 0 3 3 ...
## $ position   : Factor w/ 3 levels "D","M","S": 3 1 2 2 3 2 2 2 2 1 ...
## $ result     : Factor w/ 3 levels "D","L","W": 1 3 1 2 3 3 3 3 1 2 ...
## $ country    : Factor w/ 2 levels "England","Germany": 1 1 1 2 2 1 2 1 2 1 ...
## $ level      : Factor w/ 2 levels "1","2": 1 2 1 1 1 1 2 1 1 1 ...
```

Now our data is in a position to run a model. You may wish to conduct some exploratory data analysis at this stage similar to previous chapters, but from this chapter onward we will skip this and focus on the modeling methodology.

7.2 Modeling ordinal outcomes under the assumption of proportional odds

For simplicity, and noting that this is easily generalizable, let's assume that we have an ordinal outcome variable y with three levels similar to our walkthrough example, and that we have one input variable x. Let's call the outcome levels 1, 2 and 3. To follow our intuition from Section 7.1.1, we can model a linear continuous variable $y' = \alpha_1 x + \alpha_0 + E$, where E is some error with a mean of zero, and two increasing cutoff values τ_1 and τ_2. We define y in terms of y' as follows: $y = 1$ if $y' \leq \tau_1$, $y = 2$ if $\tau_1 < y' \leq \tau_2$ and $y = 3$ if $y' > \tau_2$.

7.2.1 Using a latent continuous outcome variable to derive a proportional odds model

Recall from Section 4.5.3 that our linear regression approach assumes that our residuals E around our line $y' = \alpha_1 x + \alpha_0$ have a normal distribution. Let's modify that assumption slightly and instead assume that our residuals take a logistic distribution based on the variance of y'. Therefore, $y' = \alpha_1 x + \alpha_0 + \sigma \epsilon$, where σ is proportional to the variance of y' and ϵ follows the shape of a logistic function. That is

$$P(\epsilon \leq z) = \frac{1}{1 + e^{-z}}$$

Let's look at the probability that our ordinal outcome variable y is in its lowest category.

$$P(y = 1) = P(y' \leq \tau_1)$$
$$= P(\alpha_1 x + \alpha_0 + \sigma \epsilon \leq \tau_1)$$
$$= P(\epsilon \leq \frac{\tau_1 - \alpha_1 x - \alpha_0}{\sigma})$$
$$= P(\epsilon \leq \gamma_1 - \beta x)$$
$$= \frac{1}{1 + e^{-(\gamma_1 - \beta x)}}$$

where $\gamma_1 = \frac{\tau_1 - \alpha_0}{\sigma}$ and $\beta = \frac{\alpha_1}{\sigma}$.

Since our only values for y are 1, 2 and 3, similar to our derivations in Section 5.2, we conclude that $P(y > 1) = 1 - P(y = 1)$, which calculates to

$$P(y > 1) = \frac{e^{-(\gamma_1 - \beta x)}}{1 + e^{-(\gamma_1 - \beta x)}}$$

Therefore

$$\frac{P(y = 1)}{P(y > 1)} = \frac{\frac{1}{1 + e^{-(\gamma_1 - \beta x)}}}{\frac{e^{-(\gamma_1 - \beta x)}}{1 + e^{-(\gamma_1 - \beta x)}}} = e^{\gamma_1 - \beta x}$$

By applying the natural logarithm, we conclude that the log odds of y being in our bottom category is

$$\ln\left(\frac{P(y = 1)}{P(y > 1)}\right) = \gamma_1 - \beta x$$

In a similar way we can derive the log odds of our ordinal outcome being in our bottom two categories as

$$\ln\left(\frac{P(y \leq 2)}{P(y = 3)}\right) = \gamma_2 - \beta x$$

where $\gamma_2 = \frac{\tau_2 - \alpha_0}{\sigma}$. One can easily see how this generalizes to an arbitrary number of ordinal categories, where we can state the log odds of being in category k or lower as

$$\ln\left(\frac{P(y \leq k)}{P(y > k)}\right) = \gamma_k - \beta x$$

Alternatively, we can state the log odds of being in a category higher than k by simply inverting the above expression:

$$\ln\left(\frac{P(y > k)}{P(y \leq k)}\right) = -(\gamma_k - \beta x) = \beta x - \gamma_k$$

By taking exponents we see that the impact of a unit change in x on the odds of y being in a higher ordinal category is β, *irrespective of what category we are looking at.* Therefore we have a single coefficient to explain the effect of x on y throughout the ordinal scale. Note that there are still different intercept coefficients γ_1 and γ_2 for each level of the ordinal scale.

7.2.2 Running a proportional odds logistic regression model

The MASS package provides a function polr() for running a proportional odds logistic regression model on a data set in a similar way to our previous models. The key (and obvious) requirement is that the outcome is an ordered factor. Since we did our conversions in Section 7.1.3 we are ready to run this model. We will start by running it on all input variables and let the polr() function handle our dummy variables automatically.

```
# run proportional odds model
library(MASS)
model <- polr(
   formula = discipline ~ n_yellow_25 + n_red_25 + position +
      country + level + result,
   data = soccer
)

# get summary
summary(model)
```

```
## Call:
## polr(formula = discipline ~ n_yellow_25 + n_red_25 + position +
##      country + level + result, data = soccer)
##
## Coefficients:
##                    Value Std. Error t value
## n_yellow_25       0.32236    0.03308  9.7456
## n_red_25          0.38324    0.04051  9.4616
## positionM         0.19685    0.11649  1.6899
## positionS        -0.68534    0.15011 -4.5655
## countryGermany    0.13297    0.09360  1.4206
## level2            0.09097    0.09355  0.9724
## resultL           0.48303    0.11195  4.3147
## resultW          -0.73947    0.12129 -6.0966
##
## Intercepts:
```

```
##              Value   Std. Error t value
## None|Yellow  2.5085  0.1918     13.0770
## Yellow|Red   3.9257  0.2057     19.0834
##
## Residual Deviance: 3444.534
## AIC: 3464.534
```

We can see that the summary returns a single set of coefficients on our input variables as we expect, with standard errors and t-statistics. We also see that there are separate intercepts for the various levels of our outcomes, as we also expect. In interpreting our model, we generally don't have a great deal of interest in the intercepts, but we will focus on the coefficients. First we would like to obtain p-values, so we can add a p-value column using the conversion methods from the t-statistic which we learned in Section 3.3.1[2].

```
# get coefficients (it's in matrix form)
coefficients <- summary(model)$coefficients

# calculate p-values
p_value <- (1 - pnorm(abs(coefficients[ ,"t value"]), 0, 1))*2

# bind back to coefficients
(coefficients <- cbind(coefficients, p_value))
```

```
##                     Value Std. Error     t value      p_value
## n_yellow_25    0.32236030 0.03307768   9.7455529 0.000000e+00
## n_red_25       0.38324333 0.04050515   9.4615947 0.000000e+00
## positionM      0.19684666 0.11648690   1.6898610 9.105456e-02
## positionS     -0.68533697 0.15011194  -4.5655060 4.982908e-06
## countryGermany 0.13297173 0.09359946   1.4206464 1.554196e-01
## level2         0.09096627 0.09354717   0.9724108 3.308462e-01
## resultL        0.48303227 0.11195131   4.3146639 1.598459e-05
## resultW       -0.73947295 0.12129301  -6.0965834 1.083595e-09
## None|Yellow    2.50850778 0.19182628  13.0769766 0.000000e+00
## Yellow|Red     3.92572124 0.20571423  19.0833721 0.000000e+00
```

Next we can convert our coefficients to odds ratios.

---

[2]Note this is not totally necessary, as significance can be estimated from viewing the confidence intervals that are formed from two standard errors either side of the coefficient estimate. However, we show how to calculate p-values here for precision purposes.

```
# calculate odds ratios
odds_ratio <- exp(coefficients[ ,"Value"])
```

We can display all our critical statistics by combining them into a dataframe.

```
# combine with coefficient and p_value
(coefficients <- cbind(
  coefficients[ ,c("Value", "p_value")],
  odds_ratio
))
```

```
##                    Value     p_value odds_ratio
## n_yellow_25    0.32236030 0.000000e+00  1.3803820
## n_red_25       0.38324333 0.000000e+00  1.4670350
## positionM      0.19684666 9.105456e-02  1.2175573
## positionS     -0.68533697 4.982908e-06  0.5039204
## countryGermany 0.13297173 1.554196e-01  1.1422177
## level2         0.09096627 3.308462e-01  1.0952321
## resultL        0.48303227 1.598459e-05  1.6209822
## resultW       -0.73947295 1.083595e-09  0.4773654
## None|Yellow    2.50850778 0.000000e+00 12.2865822
## Yellow|Red     3.92572124 0.000000e+00 50.6896241
```

Taking into consideration the p-values, we can interpret our coefficients as follows, in each case assuming that other coefficients are held still:

- Each additional yellow card received in the prior 25 games is associated with an approximately 38% higher odds of greater disciplinary action by the referee.
- Each additional red card received in the prior 25 games is associated with an approximately 47% higher odds of greater disciplinary action by the referee.
- Strikers have approximately 50% lower odds of greater disciplinary action from referees compared to Defenders.
- A player on a team that lost the game has approximately 62% higher odds of greater disciplinary action versus a player on a team that drew the game.
- A player on a team that won the game has approximately 52% lower odds of greater disciplinary action versus a player on a team that drew the game.

We can, as per previous chapters, remove the level and country variables from this model to simplify it if we wish. An examination of the coefficients and the AIC of the simpler model will reveal no substantial difference, and therefore we proceed with this model.

7.2.3 Calculating the likelihood of an observation being in a specific ordinal category

Recall from Section 7.2.1 that our proportional odds model generates multiple stratified binomial models, each of which has following form:

$$P(y \leq k) = P(y' \leq \tau_k)$$

Note that for an ordinal variable y, if $y \leq k$ and $y > k - 1$, then $y = k$. Therefore $P(y = k) = P(y \leq k) - P(y \leq k - 1)$. This means we can calculate the specific probability of an observation being in each level of the ordinal variable in our fitted model by simply calculating the difference between the fitted values from each pair of adjacent stratified binomial models. In our walkthrough example, this means we can calculate the specific probability of no action from the referee, or a yellow card being awarded, or a red card being awarded. These can be viewed using the `fitted()` function.

```
head(fitted(model))
```

```
##            None       Yellow           Red
## 1 0.8207093 0.12900184 0.05028889
## 2 0.8514232 0.10799553 0.04058128
## 3 0.7830785 0.15400189 0.06291964
## 4 0.6609864 0.22844107 0.11057249
## 5 0.9591298 0.03064719 0.01022301
## 6 0.7887766 0.15027145 0.06095200
```

It can be seen from this output how ordinal logistic regression models can be used in predictive analytics by classifying new observations into the ordinal category with the highest fitted probability. This also allows us to graphically understand the output of a proportional odds model. Figure 7.2 shows the output from a simpler proportional odds model fitted against the `n_yellow_25` and `n_red_25` input variables, with the fitted probabilities of each level of discipline from the referee plotted on the different colored surfaces. We can see in most situations that no discipline is the most likely outcome and a red card is the least likely outcome. Only at the upper ends of the scales do we see the likelihood of discipline overcoming the likelihood of no discipline, with a strong likelihood of red cards for those with an extremely poor recent disciplinary record.

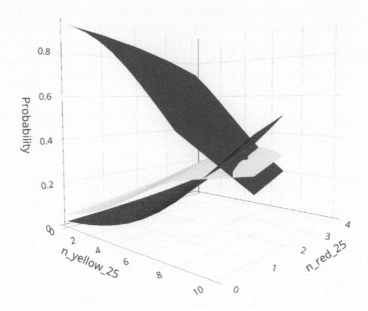

FIGURE 7.2: 3D visualization of a simple proportional odds model for `discipline` fitted against `n_yellow_25` and `n_red_25` in the `soccer` data set. Blue represents the probability of no discipline from the referee. Yellow and red represent the probability of a yellow card and a red card, respectively.

7.2.4   Model diagnostics

Similar to binomial and multinomial models, pseudo-R^2 methods are available for assessing model fit, and AIC can be used to assess model parsimony. Note that `DescTools::PseudoR2()` also offers AIC.

```
# diagnostics of simpler model
DescTools::PseudoR2(
  model,
  which = c("McFadden", "CoxSnell", "Nagelkerke", "AIC")
)
```

```
##     McFadden    CoxSnell   Nagelkerke          AIC
##    0.1009411   0.1553264    0.1912445 3464.5339371
```

There are numerous tests of goodness-of-fit that can apply to ordinal logistic regression models, and this area is the subject of considerable recent research.

The generalhoslem package in R contains routes to four possible tests, with two of them particularly recommended for ordinal models. Each work in a similar way to the Hosmer-Lemeshow test discussed in Section 5.3.2, by dividing the sample into groups and comparing the observed versus the fitted outcomes using a chi-square test. Since the null hypothesis is a good model fit, low p-values indicate potential problems with the model. We run these tests below for reference. For more information, see Fagerland and Hosmer (2017), and for a really intensive treatment of ordinal data modeling Agresti (2010) is recommended.

```
# lipsitz test
generalhoslem::lipsitz.test(model)
```

```
##
##  Lipsitz goodness of fit test for ordinal response models
##
## data: formula: discipline ~ n_yellow_25 + n_red_25 + position + country + level + formula:   result
## LR statistic = 10.429, df = 9, p-value = 0.3169
```

```
# pulkstenis-robinson test
# (requires the vector of categorical input variables as an argument)
generalhoslem::pulkrob.chisq(model, catvars = cats)
```

```
##
##  Pulkstenis-Robinson chi-squared test
##
## data: formula: discipline ~ n_yellow_25 + n_red_25 + position + country + level + formula:   result
## X-squared = 129.29, df = 137, p-value = 0.668
```

7.3 Testing the proportional odds assumption

As we discussed earlier, the suitability of a proportional odds logistic regression model depends on the assumption that each input variable has a similar effect on the different levels of the ordinal outcome variable. It is very important to check that this assumption is not violated before proceeding to declare the results of a proportional odds model valid. There are two common approaches to validating the proportional odds assumption, and we will go through each of them here.

7.3.1 Sighting the coefficients of stratified binomial models

As we learned above, proportional odds regression models effectively act as a series of stratified binomial models under the assumption that the 'slope' of the logistic function of each stratified model is the same. We can verify this by actually running stratified binomial models on our data and checking for similar coefficients on our input variables. Let's use our walkthrough example to illustrate.

Let's create two columns with binary values to correspond to the two higher levels of our ordinal variable.

```r
# create binary variable for "Yellow" or "Red" versus "None"
soccer$yellow_plus <- ifelse(soccer$discipline == "None", 0, 1)

# create binary variable for "Red" versus "Yellow" or "None"
soccer$red <- ifelse(soccer$discipline == "Red", 1, 0)
```

Now let's create two binomial logistic regression models for the two higher levels of our outcome variable.

```r
# model for at least a yellow card
yellowplus_model <- glm(
  yellow_plus ~ n_yellow_25 + n_red_25 + position +
    result + country + level,
  data = soccer,
  family = "binomial"
)

# model for a red card
red_model <- glm(
  red ~ n_yellow_25 + n_red_25 + position +
    result + country + level,
  data = soccer,
  family = "binomial"
)
```

We can now display the coefficients of both models and examine the difference between them.

```
(coefficient_comparison <- data.frame(
  yellowplus = summary(yellowplus_model)$coefficients[ , "Estimate"],
  red = summary(red_model)$coefficients[ ,"Estimate"],
  diff = summary(red_model)$coefficients[ ,"Estimate"] -
    summary(yellowplus_model)$coefficients[ , "Estimate"]
))
```

```
##                    yellowplus         red        diff
## (Intercept)       -2.63646519 -3.89865929 -1.26219410
## n_yellow_25        0.34585921  0.32468746 -0.02117176
## n_red_25           0.41454059  0.34213238 -0.07240822
## positionM          0.26108978  0.06387813 -0.19721165
## positionS         -0.72118538 -0.44228286  0.27890252
## resultL            0.46162324  0.64295195  0.18132871
## resultW           -0.77821530 -0.58536482  0.19285048
## countryGermany     0.13136665  0.10796418 -0.02340247
## level2             0.08056718  0.12421593  0.04364875
```

Ignoring the intercept, which is not of concern here, the differences appear relatively small. Large differences in coefficients would indicate that the proportional odds assumption is likely violated and alternative approaches to the problem should be considered.

7.3.2 The Brant-Wald test

In the previous method, some judgment is required to decide whether the coefficients of the stratified binomial models are 'different enough' to decide on violation of the proportional odds assumption. For those requiring more formal support, an option is the Brant-Wald test. Under this test, a generalized ordinal logistic regression model is approximated and compared to the calculated proportional odds model. A generalized ordinal logistic regression model is simply a relaxing of the proportional odds model to allow for different coefficients at each level of the ordinal outcome variable.

The Wald test is conducted on the comparison of the proportional odds and generalized models. A Wald test is a hypothesis test of the significance of the difference in model coefficients, producing a chi-square statistic. A low p-value in a Brant-Wald test is an indicator that the coefficient does not satisfy the proportional odds assumption. The brant package in R provides an implementation of the Brant-Wald test, and in this case supports our judgment that the proportional odds assumption holds.

```
library(brant)
brant::brant(model)
```

```
## ----------------------------------------------
## Test for X2  df  probability
## ----------------------------------------------
## Omnibus        14.16   8    0.08
## n_yellow_25  0.24    1    0.62
## n_red_25 1.83    1    0.18
## positionM    1.7 1    0.19
## positionS    2.33    1    0.13
## countryGermany    0.04    1    0.85
## level2         0.13    1    0.72
## resultL        1.53    1    0.22
## resultW        1.3 1    0.25
## ----------------------------------------------
##
## H0: Parallel Regression Assumption holds
```

A p-value of less than 0.05 on this test—particularly on the Omnibus plus at least one of the variables—should be interpreted as a failure of the proportional odds assumption.

7.3.3 Alternatives to proportional odds models

The proportional odds model is by far the most utilized approach to modeling ordinal outcomes (not least because of neglect in the testing of the underlying assumptions). But as we have learned, it is not always an appropriate model choice for ordinal outcomes. When the test of proportional odds fails, we need to consider a strategy for remodeling the data. If only one or two variables fail the test of proportional odds, a simple option is to remove those variables. Whether or not we are comfortable doing this will depend very much on the impact on overall model fit.

In the event where the option to remove variables is unattractive, alternative models for ordinal outcomes should be considered. The most common alternatives (which we will not cover in depth here, but are explored in Agresti (2010)) are:

- *Baseline* logistic model. This model is the same as the multinomial regression model covered in the previous chapter, using the lowest ordinal value as the reference.

- *Adjacent-category* logistic model. This model compares each level of the ordinal variable to the next highest level, and it is a constrained version of the baseline logistic model. The `brglm2` package in R offers a function `bracl()` for calculating an adjacent category logistic model.
- *Continuation-ratio* logistic model. This model compares each level of the ordinal variable to all lower levels. This can be modeled using binary logistic regression techniques, but new variables need to be constructed from the data set to allow this. The R package `rms` has a function `cr.setup()` which is a utility for preparing an outcome variable for a continuation ratio model.

7.4 Learning exercises

7.4.1 Discussion questions

1. Describe what is meant by an ordinal variable.
2. Describe how an ordinal variable can be represented using a latent continuous variable.
3. Describe the series of binomial logistic regression models that are components of a proportional odds regression model. What can you say about their coefficients?
4. If y is an ordinal outcome variable with at least three levels, and if x is an input variable that has coefficient β in a proportional odds logistic regression model, describe how to interpret the odds ratio e^{β}.
5. Describe some approaches for assessing the fit and goodness-of-fit of an ordinal logistic regression model.
6. Describe how you would use stratified binomial logistic regression models to validate the key assumption for a proportional odds model.
7. Describe a statistical significance test that can support or reject the hypothesis that the proportional odds assumption holds.
8. Describe some possible options for situations where the proportional odds assumption is violated.

7.4.2 Data exercises

Load the managers data set via the peopleanalyticsdata package or download it from the internet[3]. It is a set of information of 571 managers in a sales organization and consists of the following fields:

- employee_id for each manager
- performance_group of each manager in a recent performance review: Bottom performer, Middle performer, Top performer
- yrs_employed: total length of time employed in years
- manager_hire: whether or not the individual was hired directly to be a manager (Y) or promoted to manager (N)
- test_score: score on a test given to all managers
- group_size: the number of employees in the group they are responsible for
- concern_flag: whether or not the individual has been the subject of a complaint by a member of their group
- mobile_flag: whether or not the individual works mobile (Y) or in the office (N)
- customers: the number of customer accounts the manager is responsible for
- high_hours_flag: whether or not the manager has entered unusually high hours into their timesheet in the past year
- transfers: the number of transfer requests coming from the manager's group while they have been a manager
- reduced_schedule: whether the manager works part time (Y) or full time (N)
- city: the current office of the manager.

Construct a model to determine how the data provided may help explain the performance_group of a manager by following these steps:

1. Convert the outcome variable to an ordered factor of increasing performance.
2. Convert input variables to categorical factors as appropriate.
3. Perform any exploratory data analysis that you wish to do.
4. Run a proportional odds logistic regression model against all relevant input variables.
5. Construct p-values for the coefficients and consider how to simplify the model to remove variables that do not impact the outcome.
6. Calculate the odds ratios for your simplified model and write an interpretation of them.
7. Estimate the fit of the simplified model using a variety of metrics and perform tests to determine if the model is a good fit for the data.

---

[3]http://peopleanalytics-regression-book.org/data/managers.csv

8. Construct new outcome variables and use a stratified binomial approach to determine if the proportional odds assumption holds for your simplified model. Are there any input variables for which you may be concerned that the assumption is violated? What would you consider doing in this case?

9. Use the Brant-Wald test to support or reject the hypothesis that the proportional odds assumption holds for your simplified model.

10. Write a full report on your model intended for an audience of people with limited knowledge of statistics.

8

Modeling Explicit and Latent Hierarchy in Data

So far in this book we have learned all of the most widely used and foundational regression techniques for inferential modeling. Starting with this chapter, we will look at situations where we need to adapt or combine techniques to address certain inference goals or data characteristics. In this chapter we look at some situations where data has a hierarchy and where we wish to consider this hierarchy in our modeling efforts.

It is very often the case that data has an explicit hierarchy. For example, each observation in our data may refer to a different individual and each such individual may be a member of a few different groups. Similarly, each observation might refer to an event involving an individual, and we may have data on multiple events for the same individual. For a particular problem that we are modeling, we may wish to take into consideration the effect of the hierarchical grouping. This requires a model which has a mixture of random effects and fixed effects—called a *mixed model*.

Separately, it can be the case that data we are given could have a latent hierarchy. The input variables in the data might be measures of a smaller set of higher-level latent constructs, and we may have a more interpretable model if we hypothesize, confirm and model those latent constructs against our outcome of interest rather than using a larger number of explicit input variables. Latent variable modeling is a common technique to address this situation, and in this chapter we will review a form of latent variable modeling called *structural equation modeling*, which is very effective especially in making inferences from survey instruments with large numbers of items.

These topics are quite broad, and there are many different approaches, techniques and terms involved in mixed modeling and latent variable modeling. In this chapter we will only cover some of the simpler approaches, which would suffice for the majority of common situations in people analytics. For a deeper treatment of these topics, see Jiang (2007) for mixed models and Bartholomew, Knott, and Moustaki (2011) or Skrondal and Rabe-Hesketh (2004) for latent variable models.

DOI: 10.1201/9781003194156-8

8.1 Mixed models for explicit hierarchy in data

The most common explicit hierarchies that we see in data are group-based and time-based. A group-based hierarchy occurs when we are taking observations that belong to different groups. For example, in our first walkthrough example in Chapter 4, we modeled final examination performance against examination performance for the previous three years. In this case we considered each student observation to be independent and identically distributed, and we ran a linear regression model on all the students. If we were to receive additional information that these students were actually a mix of students in different degree programs, then we may wish to take this into account in how we model the problem—that is, we would want to assume that each student observation is only independent and identically distributed within each degree program.

Similarly, a time-based hierarchy occurs when we have multiple observations of the same subject taken at different times. For example, if we are conducting a weekly survey on the same people over the course of a year, and we are modeling how answers to some questions might depend on answers to others, we may wish to consider the effect of the person on this model.

Both of these situations introduce a new grouping variable into the problem we are modeling, thus creating a hierarchy. It is not hard to imagine that analyzing each group may produce different statistical properties compared to analyzing the entire population—for example, there could be correlations between the data inside groups which are less evident when looking at the overall population. Therefore in some cases a model may provide more useful inferences if this grouping is taken into account.

8.1.1 Fixed and random effects

Let's imagine that we have a set of observations consisting of a continuous outcome variable y and input variables $x_1, x_2, \ldots, x_p$. Let's also assume that we have an additional data point for each observation where we assign it to a group G. We are asked to determine the relationship between the outcome and the input variables. One option is to develop a linear model $y = \beta_0 + \beta_1 x_1 + \cdots + \beta_p x_p + \epsilon$, ignoring the group data. In this model, we assume that the coefficients all have a *fixed effect* on the input variables—that is, they act on every observation in the same way. This may be fine if there is trust that group membership is unlikely to have any impact on the relationship being modeled, or if we are comfortable making inferences about variables at the observation level only.

If, however, there is a belief that group membership may have an effect on the relationship being modeled, and if we are interested in interpreting our model at the group and observation level, then we need to adjust our model to a mixed model for more accurate and reliable inference. The most common adjustment is a *random intercept*. In this situation, we imagine that group membership has an effect on the 'starting point' of the relationship: the intercept. Therefore, for a given observation $y = \alpha_G + \beta_0 + \beta_1 x_1 + \cdots + \beta_p x_p + \epsilon$, where α_G is a random effect with a mean of zero associated with the group that the observation is a member of. This can be restated as:

$$y = \beta_G + \beta_1 x_1 + \cdots + \beta_p x_p + \epsilon$$

where $\beta_G = \alpha_G + \beta_0$, which is a random intercept with a mean of β_0.

This model is very similar to a standard linear regression model, except instead of having a fixed intercept, we have an intercept that varies by group. Therefore, we will essentially have two 'levels' in our model: one at the observation level to describe y and one at the group level to describe β_G. For this reason mixed models are sometimes known as *multilevel models*.

It is not too difficult to see how this approach can be extended. For example, suppose that we believe the groups also have an effect on the coefficient of the input variable x_1 as well as the intercept. Then

$$y = \beta_{G0} + \beta_{G1} x_1 + \beta_2 x_2 + \cdots + \beta_p x_p$$

where β_{G0} is a random intercept with a mean of β_0, and β_{G1} is a *random slope* with a mean of β_1. In this case, a mixed model would return the estimated coefficients at the observation level and the statistics for the random effects β_{G0} and β_{G1} at the group level.

Finally, our model does not need to be linear for this to apply. This approach also extends to logistic models and other generalized linear models. For example, if y was a binary outcome variable and our model was a binomial logistic regression model, our last equation would translate to

$$\ln\left(\frac{P(y=1)}{P(y=0)}\right) = \beta_{G0} + \beta_{G1} x_1 + \beta_2 x_2 + \cdots + \beta_p x_p$$

8.1.2 Running a mixed model

Let's look at a fun and straightforward example of how mixed models can be useful. The speed_dating data set is a set of information captured during experiments with speed dating by students at Columbia University in New

York[1]. Each row represents one meeting between an individual and a partner of the opposite sex. The data contains the following fields:

- iid is an id number for the individual.
- gender is the gender of the individual with 0 as Female and 1 as Male.
- match indicates that the meeting resulted in a match.
- samerace indicates that both the individual and the partner were of the same race.
- race is the race of the individual, with race coded as follows: Black/African American=1, European/Caucasian-American=2, Latino/Hispanic American=3, Asian/Pacific Islander/Asian-American=4, Native American=5, Other=6.
- goal is the reason why the individual is participating in the event, coded as follows: Seemed like a fun night out=1, To meet new people=2, To get a date=3, Looking for a serious relationship=4, To say I did it=5, Other=6.
- dec is a binary rating from the individual as to whether they would like to see their partner again (1 is Yes and 0 is No).
- attr is the individual's rating out of 10 on the attractiveness of the partner.
- intel is the individual's rating out of 10 on the intelligence level of the partner.
- prob is the individual's rating out of 10 on whether they believe the partner will want to see them again.
- agediff is the absolute difference in the ages of the individual and the partner.

This data can be explored in numerous ways, but we will focus here on modeling options. We are interested in the binary outcome dec (the decision of the individual), and we would like to understand how it relates to the age difference, the racial similarity and the ratings on attr, intel and prob. First, let's assume that we don't care about how an individual makes up their mind about their speed date, and that we are only interested in the dynamics of speed date decisions. Then we would simply run a binomial logistic regression on our data set, ignoring iid and other grouping variables like race, goal and gender.

```
# if needed, get data
url <- "http://peopleanalytics-regression-book.org/data/speed_dating.csv"
speed_dating <- read.csv(url)
```

---

[1] I have simplified the data set, and the full version can be found at http://www.stat.col umbia.edu/~gelman/arm/examples/speed.dating/

```
# run standard binomial model
model <- glm(dec ~ agediff + samerace + attr + intel + prob,
             data = speed_dating,
             family = "binomial")

summary(model)
```

```
##
## Call:
## glm(formula = dec ~ agediff + samerace + attr + intel + prob,
##     family = "binomial", data = speed_dating)
##
## Deviance Residuals:
##     Min       1Q    Median       3Q      Max
## -2.6497  -0.8514   -0.3477   0.8809   2.8871
##
## Coefficients:
##               Estimate Std. Error z value Pr(>|z|)
## (Intercept) -5.812900   0.184340 -31.534   <2e-16 ***
## agediff     -0.010518   0.009029  -1.165   0.2440
## samerace    -0.093422   0.055710  -1.677   0.0936 .
## attr         0.661139   0.019382  34.111   <2e-16 ***
## intel       -0.004485   0.020763  -0.216   0.8290
## prob         0.270553   0.014565  18.575   <2e-16 ***
## ---
## Signif. codes:  0 '***' 0.001 '**' 0.01 '*' 0.05 '.' 0.1 ' ' 1
##
## (Dispersion parameter for binomial family taken to be 1)
##
##     Null deviance: 10647.3  on 7788  degrees of freedom
## Residual deviance:  8082.9  on 7783  degrees of freedom
##   (589 observations deleted due to missingness)
## AIC: 8094.9
##
## Number of Fisher Scoring iterations: 5
```

In general, we see that the factors which significantly influence the speed dating decision seem to be the attractiveness of the partner and the feeling of reciprocation of interest from the partner, and that age difference, racial similarity and intelligence do not seem to play a significant role at the level of the speed date itself.

Now let's say that we are interested in how a given *individual* weighs up these factors in coming to a decision. Different individuals may have different ingoing criteria for making speed dating decisions. As a result, each individual may have varying base likelihoods of a positive decision, and each individual may be affected by the input variables in different ways as they come to their decision. Therefore we will need to assign random effects for individuals based on iid. The lme4 package in R contains functions for performing mixed linear regression models and mixed generalized linear regression models. These functions take formulas with additional terms to define the random effects to be estimated. The function for a linear model is lmer() and for a generalized linear model is glmer().

In the simple case, let's assume that each individual has a different ingoing base likelihood of making a positive decision on a speed date. We will therefore model a random intercept according to the iid of the individual. Here we would use the formula dec ~ agediff + samerace + attr + intel + prob + (1 | iid), where (1 | iid) means 'a random effect for iid on the intercept of the model.'

```
# run binomial mixed effects model
library(lme4)

iid_intercept_model <- lme4:::glmer(
  dec ~ agediff + samerace + attr + intel + prob + (1 | iid),
  data = speed_dating,
  family = "binomial"
)


# view summary without correlation table of fixed effects
summary(iid_intercept_model,
        correlation = FALSE)


## Generalized linear mixed model fit by maximum likelihood  ['']
##   Family: binomial  ( logit )
## Formula: dec ~ agediff + samerace + attr + intel + prob + (1 | iid)
##    Data: speed_dating
##
##      AIC      BIC   logLik deviance df.resid
##   6420.3   6469.0  -3203.1   6406.3     7782
##
## Scaled residuals:
##     Min       1Q   Median       3Q      Max
## -25.6965  -0.3644  -0.0606   0.3608  25.0368
##
## Random effects:
```

```
##   Groups Name        Variance Std.Dev.
##   iid    (Intercept) 5.18     2.276
## Number of obs: 7789, groups:  iid, 541
##
## Fixed effects:
##               Estimate Std. Error z value Pr(>|z|)
## (Intercept) -12.88882     0.42143 -30.583  < 2e-16 ***
## agediff      -0.03671     0.01401  -2.621  0.00877 **
## samerace      0.20187     0.08139   2.480  0.01313 *
## attr          1.07894     0.03334  32.363  < 2e-16 ***
## intel         0.31592     0.03473   9.098  < 2e-16 ***
## prob          0.61998     0.02873  21.581  < 2e-16 ***
## ---
## Signif. codes:  0 '***' 0.001 '**' 0.01 '*' 0.05 '.' 0.1 ' ' 1
```

We can see the two levels of results in this summary. The fixed effects level gives the the coefficients of the model at an observation (speed date) level, and the random effects tell us how the intercept (or base likelihood) of that model can vary according to the individual. We see that there is considerable variance in the intercept from individual to individual, and taking this into account, we now see that the decision of an individual on a given date is significantly influenced by all the factors in this model. If we had stuck with the simple binomial model, the effects of age difference, racial similarity and intelligence at an individual level would have gotten lost, and we could have reached the erroneous conclusion that none of these really matter in speed dating.

To illustrate this graphically, Figure 8.1 shows the speed_dating data for a subset of the three individuals with IIDs 252, 254 and 256. The curve represents a plain binomial logistic regression model fitted on the attr and prob input variables, irrelevant of the IID of the individual.

Figure 8.2 shows the three separate curves for each IID generated by a mixed binomial logistic regression model with a random intercept fitted on the same two input variables. Here, we can see that different individuals process the two inputs in their decision making in different ways, leading to different individual formulas which determine the likelihood of a positive decision. While a plain binomial regression model will find the best single formula from the data irrelevant of the individual, a mixed model allows us to take these different individual formulas into account in determining the effects of the input variables.

If we believe that different individuals are influenced differently by one or more of the various decision factors they consider during a speed date, we can extend our random effects to the slope coefficients of our model. For example we could use (1 + agediff | iid) to model a random effect of iid on the

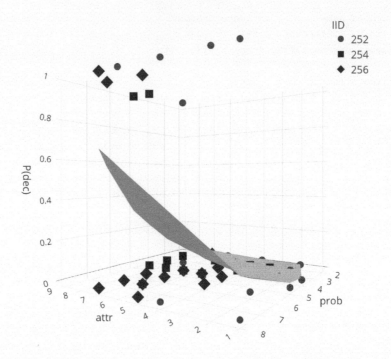

FIGURE 8.1: 3D visualization of the fitted plain binomial `model` against a subset of the `speed_dating` data for three specific `iid`s

intercept and the `agediff` coefficient. Similarly, if we wanted to consider two grouping variables—like `iid` and `goal`—on the intercept, we could add both (`1 | iid`) and (`1 | goal`) to our model formula.

8.2 Structural equation models for latent hierarchy in data

In this section we will focus entirely on survey data use cases, as this is the most common application of structural equation modeling in people analytics. However it should be noted that survey data is not the only situation where latent variables may be modeled, and this technology has substantially broader applications. Indeed, advanced practitioners may see opportunities to experiment with this technology in other use cases.

It is a frequent occurrence with surveys conducted on large samples of people, such as a public survey or a large company survey, that attempts to

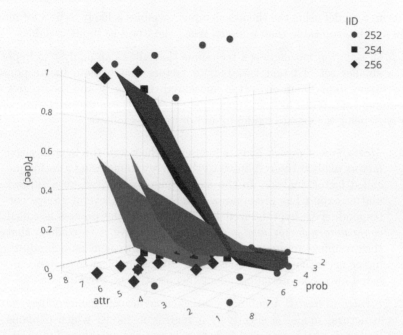

FIGURE 8.2: 3D visualization of the individual-level binomial models created by `iid_intercept_model` against a subset of the `speed_dating` data for three specific `iid`s

run regression models can be problematic due to the large number of survey questions or items. Often many of the items are highly correlated, and even if they were not, high dimensionality makes interpretability very challenging. Decision-makers are not usually interested in explanations that involve 50 or 100 variables.

Usually, such a large number of survey items are not each independently measuring a different construct. Many of the items can be considered to be addressing similar thematic constructs. For example, the items 'I believe I am compensated well' and 'I am happy with the benefits offered by my employer' could both be considered to be related to employee rewards. In some cases, survey instruments can be explicitly constructed around these themes, and in other cases, surveys have grown organically over time to include a disorganized set of items that could be grouped into themes after the fact.

It is a common request for an analyst to model a certain outcome using the many items in a complex survey as input variables. In some cases the outcome being modeled is an item in the survey itself—usually some overall measure of sentiment—or in other cases the outcome could be independent of the survey instrument, for example future attrition from the organization. In this situation, a model using the themes as input variables is likely to be a lot more useful and interpretable than a model using the items as input variables.

Structural equation modeling is a technique that allows an analyst to hypothesize a smaller set of latent variables or factors that explain the responses to the survey items themselves (the 'measured variables'), and then regresses the outcome of interest against these latent factors. It is a two-part approach, each part being a separate model in and of itself, as follows:

1. *Measurement model*: This is focused on how well the hypothesized factors explain the responses to the survey items using a technique called factor analysis. In the most common case, where a subject matter expert has pre-organized the items into several groups corresponding to hypothesized latent variables, the process is called *confirmatory factor analysis*, and the objective is to confirm that the groupings represent a high-quality measurement model, adjusting as necessary to refine the model. In the simplest case, items are fitted into separate independent themes with no overlap.

2. *Structural model*: Assuming a satisfactory measurement model, the structural model is effectively a regression model which explains how each of the proposed factors relate to the outcome of interest.

As a walkthrough example, we will work with the `politics_survey` data set.

```
# if needed, get data
url <- "http://peopleanalytics-regression-book.org/data/politics_survey.csv"
politics_survey <- read.csv(url)
```

This data set represents the results of a survey conducted by a political party on a set of approximately 2100 voters. The results are on a Likert scale of 1 to 4 where 1 indicates strong negative sentiment with a statement and 4 indicates strong positive sentiment. Subject matter experts have already grouped the items into proposed latent variables or factors, and the data takes the following form:

1. `Overall` represents the overall intention to vote for the party in the next election.
2. Items beginning with `Pol` are considered to be related to the policies of the political party.
3. Items beginning with `Hab` are considered to be related to prior voting habits in relation to the political party.
4. Items beginning with `Loc` are considered to be related to interest in local issues around where the respondent resided.
5. Items beginning with `Env` are considered to be related to interest in environmental issues.

6. Items beginning with Int are considered to be related to interest in international issues.
7. Items beginning with Pers are considered to be related to the personalities of the party representatives/leaders.
8. Items beginning with Nat are considered to be related to interest in national issues.
9. Items beginning with Eco are considered to be related to interest in economic issues.

This is a lot of data so let's just take a quick look at the first few rows and columns.

```
head(politics_survey[ ,1:7])
```

```
##    Overall Pol1 Pol2 Pol3 Hab1 Hab2 Hab3
## 1        3    2    2    2    2    2    2
## 2        4    4    4    4    4    4    4
## 3        4    4    4    4    3    2    2
## 4        3    4    4    4    3    2    2
## 5        3    3    3.   4    4    3    3
## 6        4    3    3    4    3    2    3
```

The outcome of interest here is the Overall rating. Our first aim is to confirm that the eight factors suggested by the subject matter experts represent a satisfactory measurement model (that they reasonably explain the responses to the 22 items), adjusting or refining if needed. Assuming we can confirm a satisfactory measurement model, our second aim is to run a structural model to determine how each factor relates to the overall intention to vote for the party in the next election.

8.2.1 Running and assessing the measurement model

The proposed measurement model can be seen in Figure 8.3. In this path diagram, we see the eight latent variables or factors (circles) and how they map to the individual measured items (squares) in the survey using single headed arrows. Here we are making a simplifying assumption that each latent variable influences an independent group of survey items. The diagram also notes that the latent variables may be correlated with each other, as indicated by the double-headed arrows at the top. Dashed-line paths indicate that a specific item will be used to scale the variance of the latent factor.

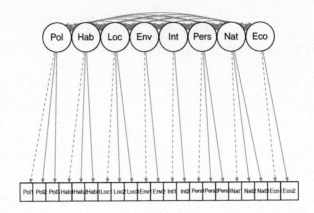

FIGURE 8.3: Simple path diagram showing proposed measurement model for `politics_survey`

The `lavaan` package in R is a specialized package for running analysis on latent variables. The function `cfa()` can be used to perform a confirmatory factor analysis on a specified measurement model. The measurement model can be specified using an appropriately commented and formatted text string as follows. Note the `=~` notation, and note also that each factor is defined on a new line.

```
# define measurement model

meas_mod <- "
# measurement model
Pol =~ Pol1 + Pol2 + Pol3
Hab =~ Hab1 + Hab2 + Hab3
Loc =~ Loc1 + Loc2 + Loc3
Env =~ Env1 + Env2
Int =~ Int1 + Int2
Pers =~ Pers1 + Pers2 + Pers3
Nat =~ Nat1 + Nat2 + Nat3
Eco =~ Eco1 + Eco2
"
```

With the measurement model defined, the confirmatory factor analysis can be run and a summary viewed. The `lavaan` summary functions used in this section produce quite large outputs spanning several pages. We will proceed to highlight which parts of this output are important in interpreting and refining the model.

```
library(lavaan)

cfa_meas_mod <- lavaan::cfa(model = meas_mod, data = politics_survey)
lavaan::summary(cfa_meas_mod, fit.measures = TRUE, standardized = TRUE)
```

```
## lavaan 0.6-7 ended normally after 108 iterations
##
##   Estimator                                        ML
##   Optimization method                          NLMINB
##   Number of free parameters                        70
##
##   Number of observations                         2108
##
## Model Test User Model:
##
##   Test statistic                              838.914
##   Degrees of freedom                              161
##   P-value (Chi-square)                          0.000
##
## Model Test Baseline Model:
##
##   Test statistic                            17137.996
##   Degrees of freedom                              210
##   P-value                                       0.000
##
## User Model versus Baseline Model:
##
##   Comparative Fit Index (CFI)                   0.960
##   Tucker-Lewis Index (TLI)                      0.948
##
## Loglikelihood and Information Criteria:
##
##   Loglikelihood user model (H0)            -37861.518
##   Loglikelihood unrestricted model (H1)    -37442.061
##
##   Akaike (AIC)                              75863.036
##   Bayesian (BIC)                            76258.780
##   Sample-size adjusted Bayesian (BIC)       76036.383
##
```

```
## Root Mean Square Error of Approximation:
##
##   RMSEA                                           0.045
##   90 Percent confidence interval - lower          0.042
##   90 Percent confidence interval - upper          0.048
##   P-value RMSEA <= 0.05                            0.998
##
## Standardized Root Mean Square Residual:
##
##   SRMR                                            0.035
##
## Parameter Estimates:
##
##   Standard errors                              Standard
##   Information                                  Expected
##   Information saturated (h1) model           Structured
##
## Latent Variables:
##                  Estimate  Std.Err  z-value  P(>|z|)   Std.lv  Std.all
##   Pol =~
##     Pol1           1.000                                0.568    0.772
##     Pol2           0.883    0.032   27.431    0.000     0.501    0.737
##     Pol3           0.488    0.024   20.575    0.000     0.277    0.512
##   Hab =~
##     Hab1           1.000                                0.623    0.755
##     Hab2           1.207    0.032   37.980    0.000     0.752    0.887
##     Hab3           1.138    0.031   36.603    0.000     0.710    0.815
##   Loc =~
##     Loc1           1.000                                0.345    0.596
##     Loc2           1.370    0.052   26.438    0.000     0.473    0.827
##     Loc3           1.515    0.058   26.169    0.000     0.523    0.801
##   Env =~
##     Env1           1.000                                0.408    0.809
##     Env2           0.605    0.031   19.363    0.000     0.247    0.699
##   Int =~
##     Int1           1.000                                0.603    0.651
##     Int2           1.264    0.060   20.959    0.000     0.762    0.869
##   Pers =~
##     Pers1          1.000                                0.493    0.635
##     Pers2          1.048    0.041   25.793    0.000     0.517    0.770
##     Pers3          0.949    0.039   24.440    0.000     0.468    0.695
##   Nat =~
##     Nat1           1.000                                0.522    0.759
##     Nat2           0.991    0.032   31.325    0.000     0.518    0.744
##     Nat3           0.949    0.035   27.075    0.000     0.495    0.638
```

```
## Eco =~
##    Eco1           1.000                                      0.525    0.791
##    Eco2           1.094    0.042   26.243    0.000           0.575    0.743
##
## Covariances:
##                Estimate  Std.Err  z-value  P(>|z|)   Std.lv   Std.all
##    Pol ~~
##       Hab        0.165    0.011   14.947    0.000    0.466    0.466
##       Loc        0.106    0.007   15.119    0.000    0.540    0.540
##       Env        0.089    0.007   12.101    0.000    0.385    0.385
##       Int        0.146    0.012   12.248    0.000    0.425    0.425
##       Pers       0.162    0.010   15.699    0.000    0.577    0.577
##       Nat        0.177    0.010   17.209    0.000    0.596    0.596
##       Eco        0.150    0.010   15.123    0.000    0.504    0.504
##    Hab ~~
##       Loc        0.069    0.006   11.060    0.000    0.323    0.323
##       Env        0.051    0.007    7.161    0.000    0.200    0.200
##       Int        0.134    0.012   11.395    0.000    0.357    0.357
##       Pers       0.121    0.010   12.619    0.000    0.393    0.393
##       Nat        0.105    0.009   11.271    0.000    0.324    0.324
##       Eco        0.089    0.009    9.569    0.000    0.273    0.273
##    Loc ~~
##       Env        0.076    0.005   15.065    0.000    0.541    0.541
##       Int        0.091    0.007   12.192    0.000    0.438    0.438
##       Pers       0.098    0.007   14.856    0.000    0.574    0.574
##       Nat        0.116    0.007   16.780    0.000    0.642    0.642
##       Eco        0.090    0.006   14.354    0.000    0.496    0.496
##    Env ~~
##       Int        0.075    0.008    9.506    0.000    0.303    0.303
##       Pers       0.075    0.007   11.482    0.000    0.375    0.375
##       Nat        0.093    0.007   13.616    0.000    0.439    0.439
##       Eco        0.078    0.007   11.561    0.000    0.365    0.365
##    Int ~~
##       Pers       0.156    0.012   13.349    0.000    0.525    0.525
##       Nat        0.186    0.012   14.952    0.000    0.592    0.592
##       Eco        0.137    0.011   12.374    0.000    0.432    0.432
##    Pers ~~
##       Nat        0.185    0.010   17.898    0.000    0.717    0.717
##       Eco        0.153    0.010   15.945    0.000    0.590    0.590
##    Nat ~~
##       Eco        0.196    0.010   19.440    0.000    0.715    0.715
##
## Variances:
##                Estimate  Std.Err  z-value  P(>|z|)   Std.lv   Std.all
##    .Pol1         0.219    0.012   19.015    0.000    0.219    0.404
```

##	.Pol2	0.211	0.010	21.455	0.000	0.211	0.457
##	.Pol3	0.216	0.007	29.384	0.000	0.216	0.737
##	.Hab1	0.293	0.011	25.855	0.000	0.293	0.430
##	.Hab2	0.153	0.011	14.434	0.000	0.153	0.213
##	.Hab3	0.254	0.012	21.814	0.000	0.254	0.335
##	.Loc1	0.217	0.007	29.063	0.000	0.217	0.645
##	.Loc2	0.103	0.006	18.226	0.000	0.103	0.316
##	.Loc3	0.153	0.007	20.463	0.000	0.153	0.358
##	.Env1	0.088	0.008	10.643	0.000	0.088	0.345
##	.Env2	0.064	0.003	18.407	0.000	0.064	0.511
##	.Int1	0.495	0.021	23.182	0.000	0.495	0.576
##	.Int2	0.188	0.025	7.653	0.000	0.188	0.244
##	.Pers1	0.361	0.013	27.065	0.000	0.361	0.597
##	.Pers2	0.184	0.009	20.580	0.000	0.184	0.408
##	.Pers3	0.234	0.009	24.865	0.000	0.234	0.517
##	.Nat1	0.201	0.009	23.320	0.000	0.201	0.425
##	.Nat2	0.215	0.009	24.119	0.000	0.215	0.446
##	.Nat3	0.357	0.013	27.981	0.000	0.357	0.593
##	.Eco1	0.165	0.010	16.244	0.000	0.165	0.374
##	.Eco2	0.268	0.013	20.071	0.000	0.268	0.448
##	Pol	0.323	0.018	18.035	0.000	1.000	1.000
##	Hab	0.389	0.020	19.276	0.000	1.000	1.000
##	Loc	0.119	0.009	13.906	0.000	1.000	1.000
##	Env	0.166	0.011	15.560	0.000	1.000	1.000
##	Int	0.364	0.026	13.846	0.000	1.000	1.000
##	Pers	0.243	0.017	14.619	0.000	1.000	1.000
##	Nat	0.273	0.015	18.788	0.000	1.000	1.000
##	Eco	0.276	0.015	17.974	0.000	1.000	1.000

This is a large set of results, but we can focus in on some important parameters to examine. First, we note that the results did not come attached to a warning. One particular warning to look out for relates to the covariance matrix being non-positive definite. This renders some of the attempted measurement invalid and is usually caused by too small a sample size for the complexity of the measurement model. Since we did not receive this warning, we can proceed safely.

Second, we examine the fit statistics. Numerous statistics are reported[2], but for larger samples such as this data set, the following measures should be examined:

---

[2]Other reported measures include chi-square statistical tests of perfect fit for the measurement model and for the baseline model, and AIC and BIC. While these generally are not very helpful in determining the quality of a specific measurement model, they are valuable for comparing different measurement model options.

- CFI and TLI, which compare the proposed model to a baseline (null or random) model to determine if it is better. Ideally we look for both of these measures to exceed 0.95. We see that our measurement model comes very close to meeting these criteria.

- RMSEA should ideally be less than 0.06, which is met by our measurement model.

- SRMR should ideally be less than 0.08, which is met by our measurement model.

Finally, the parameter estimates for the latent variables should be examined. In particular the `Std.all` column which is similar to standardized regression coefficients. These parameters are commonly knows as factor loadings—they can be interpreted as the extent to which the item response is explained by the proposed latent variable. In general, factor loadings of 0.7 or above are considered reasonable. Factor loadings less than this may be introducing unacceptable measurement error. One option if this occurs is to drop the item completely from the measurement model, or to explore an alternative measurement model with the item assigned to another latent variable. In any case the analyst will need to balance these considerations against the need to have factors measured against multiple items wherever possible in order to minimize other aspects of measurement error.

In our case we could consider dropping `Pol3`, `Loc1`, `Pers1` and `Nat3` from the measurement model as they have factor loadings of less than 0.7 and are in factors that contain three items. We will fit this revised measurement model, and rather than printing the entire output again, we will focus here on our CFI, TLI, RMSEA and SRMR statistics to see if they have improved. It is advisable, however, that factor loadings are also checked, especially where primary items that scale the variance of latent factors have been removed.

```
meas_mod_revised <- "
# measurement model
Pol =~ Pol1 + Pol2
Hab =~ Hab1 + Hab2 + Hab3
Loc =~ Loc2 + Loc3
Env =~ Env1 + Env2
Int =~ Int1 + Int2
Pers =~ Pers2 + Pers3
Nat =~ Nat1 + Nat2
Eco =~ Eco1 + Eco2
"


cfa_meas_mod_rev <- lavaan::cfa(model = meas_mod_revised,
                                data = politics_survey)
```

```
fits <- lavaan::fitmeasures(cfa_meas_mod_rev)

fits[c("cfi", "tli", "rmsea", "srmr")]
```

```
##        cfi        tli      rmsea       srmr
## 0.97804888 0.96719394 0.03966962 0.02736629
```

We now see that our measurement model comfortably meets all fit requirements. In our case we chose to completely drop four items from the model.
Analysts may wish to experiment with relaxing criteria on dropping items, or
on reassigning items to other factors to achieve a good balance between fit
and factor measurement reliability.

8.2.2 Running and interpreting the structural model

With a satisfactory measurement model, the structural model is a simple
regression formula. The sem() function in lavaan can be used to perform a full
structural equation model including the measurement model and structural
model. Like for cfa(), an extensive output can be expected from this function,
but assuming that our measurement model is satisfactory, our key interest is
now in the structural model elements of this output.

```
# define full SEM using revised measurement model
full_sem <- "
# measurement model
Pol =~ Pol1 + Pol2
Hab =~ Hab1 + Hab2 + Hab3
Loc =~ Loc2 + Loc3
Env =~ Env1 + Env2
Int =~ Int1 + Int2
Pers =~ Pers2 + Pers3
Nat =~ Nat1 + Nat2
Eco =~ Eco1 + Eco2

# structural model
Overall ~ Pol + Hab + Loc + Env + Int + Pers + Nat + Eco
"

# run full SEM
full_model <- lavaan::sem(model = full_sem, data = politics_survey)
lavaan::summary(full_model, standardized = TRUE)
```

```
## lavaan 0.6-7 ended normally after 99 iterations
##
##    Estimator                                        ML
##    Optimization method                          NLMINB
##    Number of free parameters                        71
##
##    Number of observations                         2108
##
## Model Test User Model:
##
##    Test statistic                              465.318
##    Degrees of freedom                              100
##    P-value (Chi-square)                          0.000
##
## Parameter Estimates:
##
##    Standard errors                            Standard
##    Information                                Expected
##    Information saturated (h1) model         Structured
##
## Latent Variables:
##                   Estimate  Std.Err  z-value  P(>|z|)   Std.lv  Std.all
##    Pol =~
##      Pol1            1.000                               0.626    0.850
##      Pol2            0.714    0.029   25.038    0.000    0.447    0.657
##    Hab =~
##      Hab1            1.000                               0.630    0.763
##      Hab2            1.184    0.031   38.592    0.000    0.746    0.879
##      Hab3            1.127    0.030   37.058    0.000    0.710    0.816
##    Loc =~
##      Loc2            1.000                               0.461    0.806
##      Loc3            1.179    0.036   32.390    0.000    0.544    0.833
##    Env =~
##      Env1            1.000                               0.411    0.815
##      Env2            0.596    0.031   19.281    0.000    0.245    0.695
##    Int =~
##      Int1            1.000                               0.605    0.653
##      Int2            1.256    0.062   20.366    0.000    0.760    0.867
##    Pers =~
##      Pers2           1.000                               0.520    0.774
##      Pers3           0.939    0.036   25.818    0.000    0.488    0.726
##    Nat =~
##      Nat1            1.000                               0.511    0.742
##      Nat2            1.033    0.034   29.958    0.000    0.527    0.758
##    Eco =~
```

```
##      Eco1            1.000                                   0.529    0.797
##      Eco2            1.078    0.042   25.716   0.000         0.570    0.737
##
## Regressions:
##                   Estimate  Std.Err  z-value  P(>|z|)   Std.lv   Std.all
##      Overall ~
##        Pol           0.330    0.036    9.281   0.000         0.206    0.307
##        Hab           0.255    0.024   10.614   0.000         0.161    0.240
##        Loc           0.224    0.047    4.785   0.000         0.103    0.154
##        Env          -0.114    0.042   -2.738   0.006        -0.047   -0.070
##        Int           0.046    0.028    1.605   0.108         0.028    0.041
##        Pers          0.112    0.047    2.383   0.017         0.058    0.087
##        Nat           0.122    0.071    1.728   0.084         0.063    0.093
##        Eco           0.002    0.043    0.041   0.967         0.001    0.001
##
## Covariances:
##                   Estimate  Std.Err  z-value  P(>|z|)   Std.lv   Std.all
##      Pol ~~
##        Hab           0.183    0.012   15.476   0.000         0.465    0.465
##        Loc           0.156    0.009   16.997   0.000         0.540    0.540
##        Env           0.096    0.008   12.195   0.000         0.374    0.374
##        Int           0.162    0.013   12.523   0.000         0.427    0.427
##        Pers          0.171    0.011   15.975   0.000         0.525    0.525
##        Nat           0.195    0.011   17.798   0.000         0.610    0.610
##        Eco           0.167    0.011   15.752   0.000         0.506    0.506
##      Hab ~~
##        Loc           0.091    0.008   11.218   0.000         0.315    0.315
##        Env           0.052    0.007    7.199   0.000         0.200    0.200
##        Int           0.138    0.012   11.426   0.000         0.361    0.361
##        Pers          0.112    0.010   11.484   0.000         0.341    0.341
##        Nat           0.105    0.010   11.045   0.000         0.327    0.327
##        Eco           0.091    0.009    9.608   0.000         0.273    0.273
##      Loc ~~
##        Env           0.103    0.006   16.413   0.000         0.544    0.544
##        Int           0.120    0.010   12.529   0.000         0.429    0.429
##        Pers          0.130    0.008   16.209   0.000         0.542    0.542
##        Nat           0.153    0.008   18.203   0.000         0.648    0.648
##        Eco           0.117    0.008   14.985   0.000         0.479    0.479
##      Env ~~
##        Int           0.075    0.008    9.505   0.000         0.303    0.303
##        Pers          0.075    0.007   11.058   0.000         0.351    0.351
##        Nat           0.091    0.007   13.181   0.000         0.434    0.434
##        Eco           0.079    0.007   11.583   0.000         0.364    0.364
##      Int ~~
##        Pers          0.153    0.012   13.118   0.000         0.486    0.486
```

```
## Nat          0.173    0.012   14.159   0.000    0.560   0.560
## Eco          0.138    0.011   12.293   0.000    0.431   0.431
## Pers ~~
## Nat          0.192    0.010   18.922   0.000    0.724   0.724
## Eco          0.156    0.010   16.369   0.000    0.567   0.567
## Nat ~~
## Eco          0.192    0.010   18.968   0.000    0.710   0.710
##
## Variances:
##             Estimate  Std.Err  z-value  P(>|z|)  Std.lv  Std.all
## .Pol1         0.150    0.013   11.205   0.000    0.150   0.277
## .Pol2         0.263    0.010   25.479   0.000    0.263   0.569
## .Hab1         0.285    0.011   25.570   0.000    0.285   0.418
## .Hab2         0.163    0.010   15.746   0.000    0.163   0.227
## .Hab3         0.253    0.011   22.046   0.000    0.253   0.334
## .Loc2         0.114    0.006   18.168   0.000    0.114   0.350
## .Loc3         0.130    0.008   15.729   0.000    0.130   0.306
## .Env1         0.085    0.008   10.227   0.000    0.085   0.336
## .Env2         0.064    0.003   18.701   0.000    0.064   0.518
## .Int1         0.492    0.022   22.597   0.000    0.492   0.574
## .Int2         0.192    0.025    7.547   0.000    0.192   0.249
## .Pers2        0.181    0.010   17.472   0.000    0.181   0.401
## .Pers3        0.215    0.010   21.151   0.000    0.215   0.474
## .Nat1         0.213    0.009   22.690   0.000    0.213   0.450
## .Nat2         0.205    0.010   21.502   0.000    0.205   0.425
## .Eco1         0.160    0.010   15.413   0.000    0.160   0.364
## .Eco2         0.273    0.014   20.156   0.000    0.273   0.457
## .Overall      0.242    0.008   29.506   0.000    0.242   0.537
## Pol           0.392    0.020   19.235   0.000    1.000   1.000
## Hab           0.397    0.020   19.581   0.000    1.000   1.000
## Loc           0.213    0.011   19.724   0.000    1.000   1.000
## Env           0.169    0.011   15.606   0.000    1.000   1.000
## Int           0.366    0.027   13.699   0.000    1.000   1.000
## Pers          0.270    0.015   17.514   0.000    1.000   1.000
## Nat           0.261    0.015   17.787   0.000    1.000   1.000
## Eco           0.280    0.016   17.944   0.000    1.000   1.000
```

The Std.all column of the Regressions section of the output provides the fundamentals of the structural model—these are standardized estimates which can be approximately interpreted as the proportion of the variance of the outcome that is explained by each factor. Here we can make the following interpretations:

1. Policies, habit and interest in local issues represent the three strongest drivers of likelihood of voting for the party at the next

election, and explain approximately 70% of the overall variance in the outcome.

2. Interest in national or international issues, and interest in the economy each have no significant relationship with likelihood to vote for the party at the next election.

3. Interest in the environment has a significant negative relationship with likelihood to vote for the party at the next election.

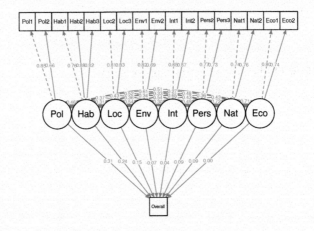

FIGURE 8.4: Path diagram for full structural equation model on `politics_survey`

The full structural equation model can be seen in Figure 8.4. This simple example illustrates the value of structural equation modeling in both reducing the dimensions of a complex regression problem and in developing more intuitive and interpretable results for stakeholders. The underlying theory of latent variable modeling, and its implementation in the lavaan package, offer much more flexibility and parameter control options than illustrated here and further exploration is highly recommended. Bartholomew, Knott, and Moustaki (2011) and Skrondal and Rabe-Hesketh (2004) are excellent resources for a deeper study of the theory and a wider range of case examples.

8.3 Learning exercises

8.3.1 Discussion questions

1. Describe some common forms of explicit hierarchies in data. Can you think of some data sets that you have worked with recently that contain an explicit hierarchy?
2. Describe the meaning of 'fixed effect' and 'random effect' in a mixed regression model.
3. Which parameter in a mixed regression model is most commonly used when applying a random effect?
4. Describe why mixed models are sometimes referred to as multilevel models.
5. In a two-level mixed model, describe the two levels of statistics that are produced and how to interpret these statistics.
6. In latent variable modeling, what is the difference between a latent variable and a measured variable?
7. Describe some reasons why latent variable modeling can be valuable in practice.
8. Describe the two components of a structural equation model. What is the purpose of each component?
9. What are the steps involved in a confirmatory factor analysis on a sufficiently large data set? Describe some fit criteria and the ideal standards for those criteria.
10. Describe a process for refining a factor analysis based on fit criteria and factor loadings. What considerations should be addressed during this process?

8.3.2 Data exercises

For Exercises 1–4, use the `speed_dating` set used earlier in this chapter[3].

1. Split the data into two sets according to the gender of the participant. Run standard binomial logistic regression models on each set to determine the relationship between the `dec` decision outcome and the input variables `samerace`, `agediff`, `attr`, `intel` and `prob`.
2. Run similar mixed models on these sets with a random intercept for `iid`.
3. What different conclusions can you make in comparing the mixed models with the standard models?

---

[3]http://peopleanalytics-regression-book.org/data/speed_dating.csv

4. Experiment with some random slope effects to see if they reveal anything new about the input variables.

For exercises 5–10, load the employee_survey data set via the peopleanalyticsdata package or download it from the internet[4]. This data set contains the results of an engagement survey of employees of a technology company. Each row represents the responses of an individual to the survey and each column represents a specific survey question, with responses on a Likert scale of 1 to 4, with 1 indicating strongly negative sentiment and 4 indicating strongly positive sentiment. Subject matter experts have grouped the items into hypothesized latent factors as follows:

- Happiness is an overall measure of the employees current sentiment about their job.
- Items beginning with Ben relate to employment benefits.
- Items beginning with Work relate to the general work environment.
- Items beginning with Man relate to perceptions of management.
- Items beginning with Car relate to perceptions of career prospects.

5. Write out the proposed measurement model, defining the latent factors in terms of the measured items.
6. Run a confirmatory factor analysis on the proposed measurement model. Examine the fit and the factor loadings.
7. Experiment with the removal of measured items from the measurement model in order to improve the overall fit.
8. Once satisfied with the fit of the measurement model, run a full structural equation model on the data.
9. Interpret the results of the structural model. Which factors appear most related to overall employee sentiment? Approximately what proportion of the variance in overall sentiment does the model explain?
10. If you dropped measured items from your measurement model, experiment with assigning them to other factors to see if this improves the fit of the model. What statistics would you use to compare different measurement models?

---

[4]http://peopleanalytics-regression-book.org/data/employee_survey.csv

9

Survival Analysis for Modeling Singular Events Over Time

In previous chapters, the outcomes we have been modeling have or have not occurred at a particular point in time following when the input variables were measured. For example, in Chapter 4 input variables were measured in the first three years of an education program and the outcome was measured at the end of the fourth year. In many situations, the outcome we are interested in is a singular event that can occur at any time following when the input variables were measured, can occur at a different time for different individuals, and once it has occurred it cannot reoccur or repeat. In medical studies, death can occur or the onset of a disease can be diagnosed at any time during the study period. In employment contexts, an attrition event can occur at various times throughout the year.

An obvious and simple way to deal with this would be to simply agree to look at a specific point in time and measure whether or not the event had occurred at that point, for example, 'How many employees had left at the three-year point?' Such an approach allows us to use standard generic regression models like those studied in previous chapters. But this approach has limitations.

Firstly, we are only able to infer conclusions about the likelihood of the event having occurred as at the end of the period of study. We cannot make inferences about the likelihood of the event throughout the period of study. Being able to say that attrition is twice as likely for certain types of individuals *at any time throughout the three years* is more powerful than merely saying that attrition is twice as likely at the three-year point.

Secondly, our sample size is constrained by the state of our data at the end of the period of study. Therefore if we lose track of an individual after two years and six months, that observation needs to be dropped from our data set if we are focused only on the three-year point. Wherever possible, loss of data is something a statistician will want to avoid as it affects the accuracy and statistical power of inferences, and also means research effort was wasted.

Survival analysis is a general term for the modeling of a time-associated binary non-repeated outcome, usually involving an understanding of the comparative risk of that outcome between two or more different groups of interest. There are two common components in an elementary survival analysis, as follows:

DOI: 10.1201/9781003194156-9

- A graphical representation of the future outcome risk of the different groups over time, using *survival curves* based on Kaplan-Meier estimates of survival rate. This is usually an effective way to establish *prima facie* relevance of a certain input variable to the survival outcome and is a very effective visual way of communicating the relevance of the input variable to non-statisticians.
- A *Cox proportional hazard* regression model to establish statistical significance of input variables and to estimate the effect of each input variable on the comparative risk of the outcome throughout the study period.

Those seeking a more in depth treatment of survival analysis should consult texts on its use in medical/clinical contexts, and a recommended source is Collett (2015). In this chapter we will use a walkthrough example to illustrate a typical use of survival analysis in a people analytics context.

The `job_retention` data set shows the results of a study of around 3,800 individuals employed in various fields of employment over a one-year period. At the beginning of the study, the individuals were asked to rate their sentiment towards their job. These individuals were then followed up monthly for a year to determine if they were still working in the same job or had left their job for a substantially different job. If an individual was not successfully followed up in a given month, they were no longer followed up for the remainder of the study period.

```
# if needed, get job_retention data
url <- "http://peopleanalytics-regression-book.org/data/job_retention.csv"
job_retention <- read.csv(url)
head(job_retention)
```

```
##    gender                  field   level sentiment intention left month
## 1       M      Public/Government    High         3         8    1     1
## 2       F                Finance     Low         8         4    0    12
## 3       M Education and Training  Medium         7         7    1     5
## 4       M                Finance     Low         8         4    0    12
## 5       M                Finance    High         7         6    1     1
## 6       F                 Health  Medium         6        10    1     2
```

For this walkthrough example, the particular fields we are interested in are:

- `gender`: The gender of the individual studied
- `field`: The field of employment that they worked in at the beginning of the study
- `level`: The level of the position in their organization at the beginning of the study—Low, Medium or High

- sentiment: The sentiment score reported on a scale of 1 to 10 at the beginning of the study, with 1 indicating extremely negative sentiment and 10 indicating extremely positive sentiment
- left: A binary variable indicating whether or not the individual had left their job as at the last follow-up
- month: The month of the last follow-up

9.1 Tracking and illustrating survival rates over the study period

In our example, we are defining 'survival' as 'remaining in substantially the same job'. We can regard the starting point as month 0, and we are following up in each of months 1 through 12. For a given month i, we can define a survival rate S_i as follows

$$S_i = S_{i-1}(1 - \frac{l_i}{n_i})$$

where l_i is the number reported as left in month i, and n_i is the number still in substantially the same job after month $i - 1$, with $S_0 = 1$.

The survival package in R allows easy construction of survival rates on data in a similar format to that in our job_retention data set. A survival object is created using the Surv() function to track the survival rate at each time period.

```
library(survival)

# create survival object with event as 'left' and time as 'month'
retention <- Surv(event = job_retention$left,
                  time = job_retention$month)

# view unique values of retention
unique(retention)
```

```
## [1]  1 12+  5  2  3  6  8  4  8+  4+ 11 10  9  7+  5+  3+  7  9+ 11+ 12  10+  6+  2+  1+
```

We can see that our survival object records the month at which the individual had left their job if they are recorded as having done so in the data set. If not, the object records the last month at which there was a record of the individual, appended with a '+' to indicate that this was the last record available.

The survfit() function allows us to calculate *Kaplan-Meier estimates* of survival for different groups in the data so that we can compare them. We can do this using our usual formula notation but using a survival object as the outcome. Let's take a look at survival by gender.

```
# kaplan-meier estimates of survival by gender
kmestimate_gender <- survival::survfit(
  formula = Surv(event = left, time = month) ~ gender,
  data = job_retention
)

summary(kmestimate_gender)
```

```
## Call: survfit(formula = Surv(event = left, time = month) ~ gender,
##     data = job_retention)
##
##                 gender=F
##  time n.risk n.event survival std.err lower 95% CI upper 95% CI
##     1   1167       7    0.994 0.00226        0.990        0.998
##     2   1140      24    0.973 0.00477        0.964        0.982
##     3   1102      45    0.933 0.00739        0.919        0.948
##     4   1044      45    0.893 0.00919        0.875        0.911
##     5    987      30    0.866 0.01016        0.846        0.886
##     6    940      51    0.819 0.01154        0.797        0.842
##     7    882      43    0.779 0.01248        0.755        0.804
##     8    830      47    0.735 0.01333        0.709        0.762
##     9    770      40    0.697 0.01394        0.670        0.725
##    10    718      21    0.676 0.01422        0.649        0.705
##    11    687      57    0.620 0.01486        0.592        0.650
##    12    621      17    0.603 0.01501        0.575        0.633
##
##                 gender=M
##  time n.risk n.event survival std.err lower 95% CI upper 95% CI
##     1   2603      17    0.993 0.00158        0.990        0.997
##     2   2559      66    0.968 0.00347        0.961        0.975
##     3   2473     100    0.929 0.00508        0.919        0.939
##     4   2360      86    0.895 0.00607        0.883        0.907
##     5   2253      56    0.873 0.00660        0.860        0.886
##     6   2171     120    0.824 0.00756        0.810        0.839
##     7   2029      85    0.790 0.00812        0.774        0.806
##     8   1916     114    0.743 0.00875        0.726        0.760
```

```
##     9    1782      96    0.703 0.00918           0.685              0.721
##    10    1661      50    0.682 0.00938           0.664              0.700
##    11    1590     101    0.638 0.00972           0.620              0.658
##    12    1460      36    0.623 0.00983           0.604              0.642
```

We can see that the n.risk, n.event and survival columns for each group correspond to the n_i, l_i and S_i in our formula above and that the confidence intervals for each survival rate are given. This can be very useful if we wish to illustrate a likely effect of a given input variable on survival likelihood.

Let's imagine that we wish to determine if the sentiment of the individual had an impact on survival likelihood. We can divide our population into two (or more) groups based on their sentiment and compare their survival rates.

```
# create a new field to define high sentiment (>= 7)
job_retention$sentiment_category <- ifelse(
  job_retention$sentiment >= 7,
  "High",
  "Not High"
)
```

```
# generate survival rates by sentiment category
kmestimate_sentimentcat <- survival::survfit(
  formula = Surv(event = left, time = month) ~ sentiment_category,
  data = job_retention
)
```

```
summary(kmestimate_sentimentcat)
```

```
## Call: survfit(formula = Surv(event = left, time = month) ~ sentiment_category,
##      data = job_retention)
##
##                 sentiment_category=High
##  time n.risk n.event survival std.err lower 95% CI upper 95% CI
##    1    3225      15    0.995 0.00120        0.993        0.998
##    2    3167      62    0.976 0.00272        0.971        0.981
##    3    3075     120    0.938 0.00429        0.929        0.946
##    4    2932     102    0.905 0.00522        0.895        0.915
##    5    2802      65    0.884 0.00571        0.873        0.895
##    6    2700     144    0.837 0.00662        0.824        0.850
##    7    2532     110    0.801 0.00718        0.787        0.815
##    8    2389     140    0.754 0.00778        0.739        0.769
##    9    2222     112    0.716 0.00818        0.700        0.732
##   10    2077      56    0.696 0.00835        0.680        0.713
##   11    1994     134    0.650 0.00871        0.633        0.667
```

```
## 	12 	1827 	45 	0.634 0.00882 	0.617 	0.651
##
## 			sentiment_category=Not High
## 	time n.risk n.event survival std.err lower 95% CI upper 95% CI
## 	1 	545 	9 	0.983 0.00546 	0.973 	0.994
## 	2 	532 	28 	0.932 0.01084 	0.911 	0.953
## 	3 	500 	25 	0.885 0.01373 	0.859 	0.912
## 	4 	472 	29 	0.831 0.01618 	0.800 	0.863
## 	5 	438 	21 	0.791 0.01758 	0.757 	0.826
## 	6 	411 	27 	0.739 0.01906 	0.703 	0.777
## 	7 	379 	18 	0.704 0.01987 	0.666 	0.744
## 	8 	357 	21 	0.662 0.02065 	0.623 	0.704
## 	9 	330 	24 	0.614 0.02136 	0.574 	0.658
## 	10 	302 	15 	0.584 0.02171 	0.543 	0.628
## 	11 	283 	24 	0.534 0.02209 	0.493 	0.579
## 	12 	254 	8 	0.517 0.02218 	0.476 	0.563
```

We can see that survival seems to consistently trend higher for those with high sentiment towards their jobs. The ggsurvplot() function in the survminer package can visualize this neatly and also provide additional statistical information on the differences between the groups, as shown in Figure 9.1.

```
library(survminer)

# show survival curves with p-value estimate and confidence intervals
survminer::ggsurvplot(
  kmestimate_sentimentcat,
  pval = TRUE,
  conf.int = TRUE,
  palette = c("blue", "red"),
  linetype = c("solid", "dashed"),
  xlab = "Month",
  ylab = "Retention Rate"
)
```

This confirms that the survival difference between the two sentiment groups is statistically significant and provides a highly intuitive visualization of the effect of sentiment on retention throughout the period of study.

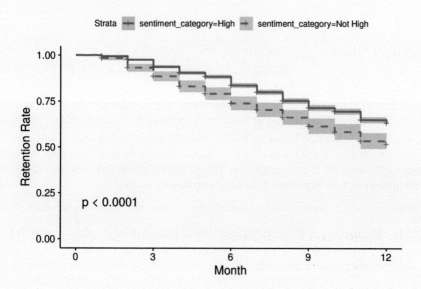

FIGURE 9.1: Survival curves by sentiment category in the `job_retention` data

---

9.2 Cox proportional hazard regression models

Let's imagine that we have a survival outcome that we are modeling for a population over a time t, and we are interested in how a set of input variables $x_1, x_2, ..., x_p$ influences that survival outcome. Given that our survival outcome is a binary variable, we can model survival at any time t as a binary logistic regression. We define $h(t)$ as the proportion who have not survived at time t, called the *hazard function*, and based on our work in Chapter 5:

$$h(t) = h_0(t)e^{\beta_1 x_1 + \beta_2 x_2 + \cdots + \beta_p x_p}$$

where $h_0(t)$ is a base or intercept hazard at time t, and β_i is the coefficient associated with x_i .

Now let's imagine we are comparing the hazard for two different individuals A and B from our population. We make an assumption that our hazard curves $h^A(t)$ for individual A and $h^B(t)$ for individual B are always proportional to each other and never cross—this is called the *proportional hazard assumption*. Under this assumption, we can conclude that

$$\frac{h^B(t)}{h^A(t)} = \frac{h_0(t)e^{\beta_1 x_1^B + \beta_2 x_2^B + \cdots + \beta_p x_p^B}}{h_0(t)e^{\beta_1 x_1^A + \beta_2 x_2^A + \cdots + \beta_p x_p^A}}$$

$$= e^{\beta_1(x_1^B - x_1^A) + \beta_2(x_2^B - x_2^A) + \ldots \beta_p(x_p^B - x_p^A)}$$

Note that there is no t in our final equation. The important observation here is that the hazard for person B relative to person A is *constant and independent of time*. This allows us to take a complicating factor out of our model. It means we can model the effect of input variables on the hazard without needing to account for changes over times, making this model very similar in interpretation to a standard binomial regression model.

9.2.1   Running a Cox proportional hazard regression model

A Cox proportional hazard model can be run using the `coxph()` function in the `survival` package, with the outcome as a survival object. Let's model our survival against the input variables `gender`, `field`, `level` and `sentiment`.

```r
# run cox model against survival outcome
cox_model <- survival::coxph(
  formula = Surv(event = left, time = month) ~ gender +
    field + level + sentiment,
  data = job_retention
)

summary(cox_model)
```

```
## Call:
## survival::coxph(formula = Surv(event = left, time = month) ~
##     gender + field + level + sentiment, data = job_retention)
##
## n= 3770, number of events= 1354
##
##                        coef exp(coef) se(coef)      z Pr(>|z|)
## genderM            -0.04548   0.95553  0.05886 -0.773 0.439647
## fieldFinance        0.22334   1.25025  0.06681  3.343 0.000829 ***
## fieldHealth         0.27830   1.32089  0.12890  2.159 0.030849 *
## fieldLaw            0.10532   1.11107  0.14515  0.726 0.468086
## fieldPublic/Government 0.11499 1.12186  0.08899  1.292 0.196277
## fieldSales/Marketing 0.08776  1.09173  0.10211  0.859 0.390082
## levelLow            0.14813   1.15967  0.09000  1.646 0.099799 .
## levelMedium         0.17666   1.19323  0.10203  1.732 0.083362 .
## sentiment          -0.11756   0.88909  0.01397 -8.415  < 2e-16 ***
```

```
## ---
## Signif. codes:  0 '***' 0.001 '**' 0.01 '*' 0.05 '.' 0.1 ' ' 1
##
##                      exp(coef) exp(-coef) lower .95 upper .95
## genderM                 0.9555     1.0465    0.8514    1.0724
## fieldFinance            1.2502     0.7998    1.0968    1.4252
## fieldHealth             1.3209     0.7571    1.0260    1.7005
## fieldLaw                1.1111     0.9000    0.8360    1.4767
## fieldPublic/Government  1.1219     0.8914    0.9423    1.3356
## fieldSales/Marketing    1.0917     0.9160    0.8937    1.3336
## levelLow                1.1597     0.8623    0.9721    1.3834
## levelMedium             1.1932     0.8381    0.9770    1.4574
## sentiment               0.8891     1.1248    0.8651    0.9138
##
## Concordance= 0.578  (se = 0.008 )
## Likelihood ratio test= 89.18  on 9 df,   p=2e-15
## Wald test           = 94.95  on 9 df,   p=<2e-16
## Score (logrank) test = 95.31  on 9 df,   p=<2e-16
```

The model returns the following[1]

- Coefficients for each input variable and their p-values. Here we can conclude that working in Finance or Health is associated with a significantly greater likelihood of leaving over the period studied, and that higher sentiment is associated with a significantly lower likelihood of leaving.

- Relative odds ratios associated with each input variable. For example, a single extra point in sentiment reduces the odds of leaving by ~11%. A single less point increases the odds of leaving by ~12%. Confidence intervals for the coefficients are also provided.

- Three statistical tests on the null hypothesis that the coefficients are zero. This null hypothesis is rejected by all three tests which can be interpreted as meaning that the model is significant.

Importantly, as well as statistically validating that sentiment has a significant effect on retention, our Cox model has allowed us to control for possible mediating variables. We can now say that sentiment has a significant effect on retention even for individuals of the same gender, in the same field and at the same level.

---

[1]The concordance measure returned is a measure of how well the model can predict in any given pair who will survive longer and is valuable in a number of medical research contexts.

9.2.2   Checking the proportional hazard assumption

Note that we mentioned in the previous section a critical assumption for our Cox proportional hazard model to be valid, called the proportional hazard assumption. As always, it is important to check this assumption before finalizing any inferences or conclusions from your model.

The most popular test of this assumption uses a residual known as a *Schoenfeld residual*, which would be expected to be independent of time if the proportional hazard assumption holds. The cox.zph() function in the survival package runs a statistical test on the null hypothesis that the Schoenfeld residuals are independent of time. The test is conducted on every input variable and on the model as a whole, and a significant result would reject the proportional hazard assumption.

```
(ph_check <- survival::cox.zph(cox_model))
```

```
##             chisq df    p
## gender      0.726  1 0.39
## field       6.656  5 0.25
## level       2.135  2 0.34
## sentiment   1.828  1 0.18
## GLOBAL     11.156  9 0.27
```

In our case, we can confirm that the proportional hazard assumption is not rejected. The ggcoxzph() function in the survminer package takes the result of the cox.zph() check and allows a graphical check by plotting the residuals against time, as seen in Figure 9.2.

```
survminer::ggcoxzph(ph_check,
                    font.main = 10,
                    font.x = 10,
                    font.y = 10)
```

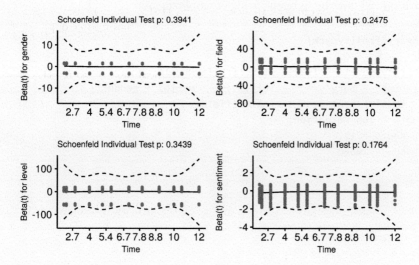

FIGURE 9.2: Schoenfeld test on proportional hazard assumption for `cox_model`

9.3 Frailty models

We noticed in our example in the previous section that certain fields of employment appeared to have a significant effect on the attrition hazard. It is therefore possible that different fields of employment have different base hazard functions, and we may wish to take this into account in determining if input variables have a significant relationship with attrition. This is analogous to a mixed model which we looked at in Section 8.1.

In this case we would apply a random intercept effect to the base hazard function $h_0(t)$ according to the field of employment of an individual, in order to take this into account in our modeling. This kind of model is called a *frailty model*, taken from the clinical context, where different groups of patients may have different frailties (background risks of death).

There are many variants of how frailty models are run in the clinical context (see Collett (2015) for an excellent exposition of these), but the main application of a frailty model in people analytics would be to adapt a Cox proportional hazard model to take into account different background risks of the hazard event occurring among different groups in the data. This is called a *shared frailty model*. The `frailtypack` R package allows various frailty models to be run with relative ease. This is how we would run a shared frailty

model on our `job_retention` data to take account of the different background
attrition risk for the different fields of employment.

```
library(frailtypack)

(frailty_model <- frailtypack::frailtyPenal(
  formula = Surv(event = left, time = month) ~ gender +
    level + sentiment + cluster(field),
  data = job_retention,
  n.knots = 12,
  kappa = 10000
))
```

```
##
## Be patient. The program is computing ...
## The program took 1.53 seconds


## Call:
## frailtypack::frailtyPenal(formula = Surv(event = left, time = month) ~
##     gender + level + sentiment + cluster(field), data = job_retention,
##     n.knots = 12, kappa = 10000)
##
##
##   Shared Gamma Frailty model parameter estimates
##   using a Penalized Likelihood on the hazard function
##
##                 coef exp(coef) SE coef (H) SE coef (HIH)        z          p
## genderM    -0.029531  0.970901   0.0591820     0.0591820 -0.498986 6.1779e-01
## levelLow    0.198548  1.219630   0.0917396     0.0917396  2.164255 3.0445e-02
## levelMedium 0.223266  1.250154   0.1035510     0.1035510  2.156101 3.1076e-02
## sentiment  -0.108262  0.897392   0.0141325     0.0141325 -7.660518 1.8541e-14
##
##         chisq df global p
## level 5.28624  2    0.0711
##
##   Frailty parameter, Theta: 48.3209 (SE (H): 25.5895 ) p = 0.029492
##
##   penalized marginal log-likelihood = -5510.36
##   Convergence criteria:
##   parameters = 3.05e-05 likelihood = 4.91e-06 gradient = 1.55e-09
##
##   LCV = the approximate likelihood cross-validation criterion
##         in the semi parametrical case    = 1.46587
##
##   n= 3770
##   n events= 1354   n groups= 6
##   number of iterations:  18
##
##   Exact number of knots used:  12
##   Value of the smoothing parameter:  10000, DoF:  6.31
```

We can see that the frailty parameter is significant, indicating that there is
sufficient difference in the background attrition risk to justify the application
of a random hazard effect. We also see that the level of employment now

becomes more significant in addition to sentiment, with Low and Medium level employees more likely to leave compared to High level employees.

The frailtyPenal() function can also be a useful way to observe the different baseline survivals for groups in the data. For example, a simple stratified Cox proportional hazard model based on sentiment category can be constructed[2].

```
stratified_base <- frailtypack::frailtyPenal(
  formula = Surv(event = left, time = month) ~
    strata(sentiment_category),
  data = job_retention,
  n.knots = 12,
  kappa = rep(10000, 2)
)
```

This can then be plotted to observe how baseline retention differs by group, as in Figure 9.3[3].

```
plot(stratified_base, type.plot = "Survival",
    pos.legend = "topright", Xlab = "Month",
    Ylab = "Baseline retention rate",
    color = 1)
```

---

[2]Note there needs to be a kappa for each level of the stratification.
[3]This is another route to calculating survival curves similar to Figure 9.1.

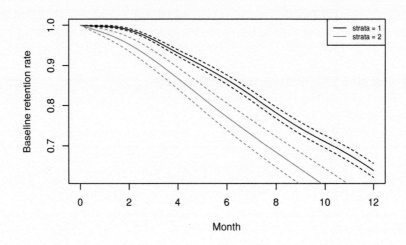

FIGURE 9.3: Baseline retention curves for the two sentiment categories in the `job_retention` data set

9.4   Learning exercises

9.4.1   Discussion questions

1. Describe some of the reasons why a survival analysis is a useful tool for analyzing data where outcome events happen at different times.
2. Describe the Kaplan-Meier survival estimate and how it is calculated.
3. What are some common uses for survival curves in practice?
4. Why is it important to run a Cox proportional hazard model in addition to calculating survival estimates when trying to understand the effect of a given variable on survival?
5. Describe the assumption that underlies a Cox proportional hazard model and how this assumption can be checked.
6. What is a frailty model, and why might it be useful in the context of survival analysis?

9.4.2 Data exercises

For these exercises, use the same `job_retention` data set as in the walkthrough example for this chapter, which can be loaded via the `peopleanalyticsdata` package or downloaded from the internet[4]. The `intention` field represents a score of 1 to 10 on the individual's intention to leave their job in the next 12 months, where 1 indicates an extremely low intention and 10 indicates an extremely high intention. This response was recorded at the beginning of the study period.

1. Create three categories of `intention` as follows: High (score of 7 or higher), Moderate (score of 4–6), Low (score of 3 or less)
2. Calculate Kaplan-Meier survival estimates for the three categories and visualize these using survival curves.
3. Determine the effect of `intention` on retention using a Cox proportional hazard model, controlling for `gender`, `field` and `level`.
4. Perform an appropriate check that the proportional hazard assumption holds for your model.
5. Run a similar model, but this time include the `sentiment` input variable. How would you interpret the results?
6. Experiment with running a frailty model to take into account the different background attrition risk by field of employment.

---

[4]http://peopleanalytics-regression-book.org/data/job_retention.csv

10

Alternative Technical Approaches in R and Python

As outlined earlier in this book, all technical implementations of the modeling techniques in previous chapters have relied wherever possible on base R code and specialist packages for specific methodologies—this allowed a focus on the basics of understanding, running and interpreting these models which is the key aim of this book. For those interested in a wider range of technical options for running inferential statistical models, this chapter illustrates some alternative options and should be considered a starting point for those interested rather than an in-depth exposition.

First we look at options for generating models in more predictable formats in R. We have seen in prior chapters that the output of many models in R can be inconsistent. In many cases we are given more information than we need, and in some cases we have less than we need. Formats can vary, and we sometimes need to look in different parts of the output to see specific statistics that we seek. The `tidymodels` set of packages tries to bring the principles of tidy data into the realm of statistical modeling and we will illustrate this briefly.

Second, for those whose preference is to use Python, we provide some examples for how inferential regression models can be run in Python. While Python is particularly well-tooled for running predictive models, it does not have the full range of statistical inference tools that are available in R. In particular, using predictive modeling or machine learning packages like `scikit-learn` to conduct regression modeling can often leave the analyst lacking when seeking information about certain model statistics when those statistics are not typically sought after in a predictive modeling workflow. We briefly illustrate some Python packages which perform modeling with a greater emphasis on inference versus prediction.

DOI: 10.1201/9781003194156-10

10.1   'Tidier' modeling approaches in R

The `tidymodels` meta-package is a collection of packages which collectively apply the principles of tidy data to the construction of statistical models. More information and learning resources on `tidymodels` can be found at `https://www.tidymodels.org/`. Within `tidymodels` there are two packages which are particularly useful in controlling the output of models in R: the `broom` and `parsnip` packages.

10.1.1   The `broom` package

Consistent with how it is named, `broom` aims to tidy up the output of the models into a predictable format. It works with over 100 different types of models in R. In order to illustrate its use, let's run a model from a previous chapter—specifically our salesperson promotion model in Chapter 5.

```
# obtain salespeople data
url <- "http://peopleanalytics-regression-book.org/data/salespeople.csv"
salespeople <- read.csv(url)
```

As in Chapter 5, we convert the `promoted` column to a factor and run a binomial logistic regression model on the `promoted` outcome.

```
# convert promoted to factor
salespeople$promoted <- as.factor(salespeople$promoted)

# build model to predict promotion based on sales and customer_rate
promotion_model <- glm(formula = promoted ~ sales + customer_rate,
                       family = "binomial",
                       data = salespeople)
```

We now have our model sitting in memory. We can use three key functions in the `broom` package to view a variety of model statistics. First, the `tidy()` function allows us to see the coefficient statistics of the model.

```
# load tidymodels metapackage
library(tidymodels)

# view coefficient statistics
broom::tidy(promotion_model)
```

```
## # A tibble: 3 x 5
##   term          estimate std.error statistic  p.value
##   <chr>            <dbl>     <dbl>     <dbl>    <dbl>
## 1 (Intercept)    -19.5      3.35      -5.83 5.48e- 9
## 2 sales            0.0404   0.00653    6.19 6.03e-10
## 3 customer_rate   -1.12     0.467     -2.40 1.63e- 2
```

The glance() function allows us to see a row of overall model statistics:

```
# view model statistics
broom::glance(promotion_model)
```

```
## # A tibble: 1 x 8
##   null.deviance df.null logLik   AIC   BIC deviance df.residual  nobs
##           <dbl>   <int>  <dbl> <dbl> <dbl>    <dbl>       <int> <int>
## 1          440.     349  -32.6  71.1  82.7     65.1         347   350
```

The augment() function augments the observations in the data set with a range of observation-level model statistics such as residuals:

```
# view augmented data
head(broom::augment(promotion_model))
```

```
## # A tibble: 6 x 9
##   promoted sales customer_rate .fitted  .resid .std.resid    .hat .sigma  .cooksd
##   <fct>    <int>         <dbl>   <dbl>   <dbl>      <dbl>   <dbl>  <dbl>    <dbl>
## 1 0          594          3.94  0.0522 -1.20      -1.22  0.0289  0.429 1.08e- 2
## 2 0          446          4.06 -6.06   -0.0683    -0.0684 0.00212 0.434 1.66e- 6
## 3 1          674          3.83  3.41    0.255      0.257  0.0161  0.434 1.84e- 4
## 4 0          525          3.62 -2.38   -0.422     -0.425  0.0153  0.433 4.90e- 4
## 5 1          657          4.4   2.08    0.485      0.493  0.0315  0.433 1.40e- 3
## 6 1          918          4.54 12.5     0.00278    0.00278 0.0000174 0.434 2.24e-11
```

These functions are model-agnostic for a very wide range of common models in R. For example, we can use them on our proportional odds model on soccer discipline from Chapter 7, and they will generate the relevant statistics in tidy tables.

```r
# get soccer data
url <- "http://peopleanalytics-regression-book.org/data/soccer.csv"
soccer <- read.csv(url)

# convert discipline to ordered factor
soccer$discipline <- ordered(soccer$discipline,
                      levels = c("None", "Yellow", "Red"))

# run proportional odds model
library(MASS)
soccer_model <- polr(
  formula = discipline ~ n_yellow_25 + n_red_25 + position +
    country + level + result,
  data = soccer
)

# view model statistics
broom::glance(soccer_model)
```

```
## # A tibble: 1 x 7
##      edf logLik   AIC   BIC deviance df.residual  nobs
##    <int>  <dbl> <dbl> <dbl>    <dbl>       <dbl> <dbl>
## 1     10 -1722. 3465. 3522.    3445.        2281  2291
```

broom functions integrate well into other tidyverse methods, and allow easy running of models over nested subsets of data. For example, if we want to run our soccer discipline model across the different countries in the data set and see all the model statistics in a neat table, we can use typical tidyverse grammar to do so using dplyr.

```
# load the tidyverse metapackage (includes dplyr)
library(tidyverse)

# define function to run soccer model and glance at results
soccer_model_glance <- function(form, df) {
  model <- polr(formula = form, data = df)
  broom::glance(model)
}

# run it nested by country
soccer %>%
  dplyr::nest_by(country) %>%
  dplyr::summarise(
    soccer_model_glance("discipline ~ n_yellow_25 + n_red_25", data)
  )
```

```
## # A tibble: 2 x 8
## # Groups:   country [2]
##    country    edf logLik    AIC   BIC deviance df.residual  nobs
##    <chr>    <int>  <dbl>  <dbl> <dbl>    <dbl>       <dbl> <dbl>
## 1 England      4  -883. 1773. 1794.    1765.        1128  1132
## 2 Germany      4  -926. 1861. 1881.    1853.        1155  1159
```

In a similar way, by putting model formulas in a dataframe column, numerous models can be run in a single command and results viewed in a tidy dataframe.

```
# create model formula column
formula <- c(
  "discipline ~ n_yellow_25",
  "discipline ~ n_yellow_25 + n_red_25",
  "discipline ~ n_yellow_25 + n_red_25 + position"
)

# create dataframe
models <- data.frame(formula)

# run models and glance at results
models %>%
  dplyr::group_by(formula) %>%
  dplyr::summarise(soccer_model_glance(formula, soccer))
```

```
## # A tibble: 3 x 8
##   formula                                          edf logLik  AIC   BIC deviance df.residual  nobs
##   <chr>                                          <int>  <dbl> <dbl> <dbl>    <dbl>       <dbl> <dbl>
## 1 discipline ~ n_yellow_25                          3 -1861. 3728. 3745.    3722.        2288  2291
## 2 discipline ~ n_yellow_25 + n_red_25               4 -1809. 3627. 3650.    3619.        2287  2291
## 3 discipline ~ n_yellow_25 + n_red_25 + position    6 -1783. 3579. 3613.    3567.        2285  2291
```

10.1.2   The `parsnip` package

The `parsnip` package aims to create a unified interface to running models, to avoid users needing to understand different model terminology and other minutiae. It also takes a more hierarchical approach to defining models that is similar in nature to the object-oriented approaches that Python users would be more familiar with.

Again let's use our salesperson promotion model example to illustrate. We start by defining a model family that we wish to use, in this case logistic regression, and define a specific engine and mode.

```
model <- parsnip::logistic_reg() %>%
  parsnip::set_engine("glm") %>%
  parsnip::set_mode("classification")
```

We can use the `translate()` function to see what kind of model we have created:

```
model %>%
  parsnip::translate()
```

```
## Logistic Regression Model Specification (classification)
##
## Computational engine: glm
##
## Model fit template:
## stats::glm(formula = missing_arg(), data = missing_arg(),
##     weights = missing_arg(), family = stats::binomial)
```

Now with our model defined, we can fit it using a formula and data and then use broom to view the coefficients:

```
model %>%
  parsnip::fit(formula = promoted ~ sales + customer_rate,
               data = salespeople) %>%
  broom::tidy()
```

```
## # A tibble: 3 x 5
##   term          estimate std.error statistic  p.value
##   <chr>            <dbl>     <dbl>     <dbl>    <dbl>
## 1 (Intercept)     -19.5      3.35     -5.83 5.48e- 9
## 2 sales            0.0404    0.00653   6.19 6.03e-10
## 3 customer_rate   -1.12      0.467    -2.40 1.63e- 2
```

parsnip functions are particularly motivated around tooling for machine learning model workflows in a similar way to scikit-learn in Python, but they can offer an attractive approach to coding inferential models, particularly where common families of models are used.

10.2 Inferential statistical modeling in Python

In general, the modeling functions contained in scikit-learn—which tends to be the go-to modeling package for most Python users—are oriented towards predictive modeling and can be challenging to navigate for those who are primarily interested in inferential modeling. In this section we will briefly review approaches for running some of the models contained in this book in Python. The statsmodels package is highly recommended as it offers a wide range of models which report similar statistics to those reviewed in this book. Full statsmodels documentation can be found at https://www.statsmodels.org/stable/index.html.

10.2.1 Ordinary Least Squares (OLS) linear regression

The OLS linear regression model reviewed in Chapter 4 can be generated using the statsmodels package, which can report a reasonably thorough set of model statistics. By using the statsmodels formula API, model formulas similar to those used in R can be used.

```python
import pandas as pd
import statsmodels.formula.api as smf

# get data
url = "http://peopleanalytics-regression-book.org/data/ugtests.csv"
ugtests = pd.read_csv(url)

# define model
model = smf.ols(formula = "Final ~ Yr3 + Yr2 + Yr1", data = ugtests)

# fit model
ugtests_model = model.fit()

# see results summary
print(ugtests_model.summary())
```

```
##                            OLS Regression Results
## ==============================================================================
## Dep. Variable:                  Final   R-squared:                       0.530
## Model:                            OLS   Adj. R-squared:                  0.529
## Method:                 Least Squares   F-statistic:                     365.5
## Date:                Mon, 19 Apr 2021   Prob (F-statistic):           8.22e-159
## Time:                        09:41:09   Log-Likelihood:                 -4711.6
## No. Observations:                 975   AIC:                             9431.
## Df Residuals:                     971   BIC:                             9451.
## Df Model:                           3
## Covariance Type:            nonrobust
## ==============================================================================
##                  coef    std err          t      P>|t|      [0.025      0.975]
## ------------------------------------------------------------------------------
## Intercept     14.1460      5.480      2.581      0.010       3.392      24.900
## Yr3            0.8657      0.029     29.710      0.000       0.809       0.923
## Yr2            0.4313      0.033     13.267      0.000       0.367       0.495
## Yr1            0.0760      0.065      1.163      0.245      -0.052       0.204
## ==============================================================================
## Omnibus:                        0.762   Durbin-Watson:                   2.006
## Prob(Omnibus):                  0.683   Jarque-Bera (JB):                0.795
## Skew:                           0.067   Prob(JB):                        0.672
## Kurtosis:                       2.961   Cond. No.                         858.
## ==============================================================================
##
## Notes:
## [1] Standard Errors assume that the covariance matrix of the errors is
##     correctly specified.
```

10.2.2   Binomial logistic regression

Binomial logistic regression models can be generated in a similar way to OLS linear regression models using the statsmodels formula API, calling the binomial family from the general statsmodels API.

```python
import pandas as pd
import statsmodels.api as sm
import statsmodels.formula.api as smf

# obtain salespeople data
url = "http://peopleanalytics-regression-book.org/data/salespeople.csv"
salespeople = pd.read_csv(url)

# define model
model = smf.glm(formula = "promoted ~ sales + customer_rate",
                data = salespeople,
                family = sm.families.Binomial())


# fit model
promotion_model = model.fit()


# see results summary
print(promotion_model.summary())
```

```
##                 Generalized Linear Model Regression Results
## ==============================================================================
## Dep. Variable:             promoted   No. Observations:              350
## Model:                          GLM   Df Residuals:                  347
## Model Family:              Binomial   Df Model:                        2
## Link Function:                logit   Scale:                      1.0000
## Method:                        IRLS   Log-Likelihood:            -32.566
## Date:              Mon, 19 Apr 2021   Deviance:                   65.131
## Time:                      09:41:09   Pearson chi2:                 198.
## No. Iterations:                   9
## Covariance Type:          nonrobust
## ==============================================================================
##                  coef    std err          z      P>|z|      [0.025      0.975]
## ------------------------------------------------------------------------------
## Intercept    -19.5177      3.347     -5.831      0.000     -26.078     -12.958
## sales          0.0404      0.007      6.189      0.000       0.028       0.053
## customer_rate  -1.1221      0.467     -2.403      0.016      -2.037      -0.207
## ==============================================================================
```

10.2.3   Multinomial logistic regression

Multinomial logistic regression is similarly available using the statsmodels formula API. As usual, care must be taken to ensure that the reference category is appropriately defined, dummy input variables need to be explicitly constructed, and a constant term must be added to ensure an intercept is calculated.

```python
import pandas as pd
import statsmodels.api as sm

# load health insurance data
url = "http://peopleanalytics-regression-book.org/data/health_insurance.csv"
health_insurance = pd.read_csv(url)

# convert product to categorical as an outcome variable
y = pd.Categorical(health_insurance['product'])

# create dummies for gender
X1 = pd.get_dummies(health_insurance['gender'], drop_first = True)

# replace back into input variables
X2 = health_insurance.drop(['product', 'gender'], axis = 1)
X = pd.concat([X1, X2], axis = 1)

# add a constant term to ensure intercept is calculated
Xc = sm.add_constant(X)

# define model
model = sm.MNLogit(y, Xc)

# fit model
insurance_model = model.fit()

# see results summary
print(insurance_model.summary())
```

```
##                        MNLogit Regression Results
## ==============================================================================
## Dep. Variable:                      y   No. Observations:             1453
## Model:                        MNLogit   Df Residuals:                 1439
## Method:                           MLE   Df Model:                       12
## Date:               Mon, 19 Apr 2021   Pseudo R-squ.:               0.5332
## Time:                        09:41:10   Log-Likelihood:             -744.68
## converged:                       True   LL-Null:                    -1595.3
## Covariance Type:            nonrobust   LLR p-value:                  0.000
## ==============================================================================
##          y=B       coef    std err          z      P>|z|      [0.025      0.975]
## ------------------------------------------------------------------------------
## const          -4.6010      0.511     -9.012      0.000     -5.602     -3.600
## Male           -2.3826      0.232    -10.251      0.000     -2.838     -1.927
## Non-binary      0.2528      1.226      0.206      0.837     -2.151      2.656
## age             0.2437      0.015     15.790      0.000      0.213      0.274
## household      -0.9677      0.069    -13.938      0.000     -1.104     -0.832
## position_level -0.4153      0.089     -4.658      0.000     -0.590     -0.241
## absent          0.0117      0.013      0.900      0.368     -0.014      0.037
## ------------------------------------------------------------------------------
##          y=C       coef    std err          z      P>|z|      [0.025      0.975]
## ------------------------------------------------------------------------------
## const         -10.2261      0.620    -16.501      0.000    -11.441     -9.011
## Male            0.0967      0.195      0.495      0.621     -0.286      0.480
## Non-binary     -1.2698      2.036     -0.624      0.533     -5.261      2.721
## age             0.2698      0.016     17.218      0.000      0.239      0.301
## household       0.2043      0.050      4.119      0.000      0.107      0.302
## position_level -0.2136      0.082     -2.597      0.009     -0.375     -0.052
## absent          0.0033      0.012      0.263      0.793     -0.021      0.028
## ==============================================================================
```

10.2.4 Structural equation models

The `semopy` package is a specialized package for the implementation of Structural Equation Models in Python, and its implementation is very similar to the `lavaan` package in R. However, its reporting is not as intuitive compared to `lavaan`. A full tutorial is available at https://semopy.com/tutorial.html. Here is an example of how to run the same model as that studied in Section 8.2 using `semopy`.

```python
import pandas as pd
from semopy import Model


# get data
url = "http://peopleanalytics-regression-book.org/data/politics_survey.csv"
politics_survey = pd.read_csv(url)


# define full measurement and structural model
measurement_model = """
# measurement model
Pol =~ Pol1 + Pol2
Hab =~ Hab1 + Hab2 + Hab3
Loc =~ Loc2 + Loc3
Env =~ Env1 + Env2
Int =~ Int1 + Int2
Pers =~ Pers2 + Pers3
Nat =~ Nat1 + Nat2
Eco =~ Eco1 + Eco2

# structural model
Overall ~ Pol + Hab + Loc + Env + Int + Pers + Nat + Eco
"""


full_model = Model(measurement_model)


# fit model to data and inspect
full_model.fit(politics_survey)
```

Then to inspect the results:

```python
# inspect the results of SEM (first few rows)
full_model.inspect().head()
```

```
##       lval op rval   Estimate    Std. Err   z-value  p-value
## 0    Pol1  ~  Pol   1.000000          -         -        -
## 1    Pol2  ~  Pol   0.713719   0.0285052   25.0382        0
## 2    Hab1  ~  Hab   1.000000          -         -        -
## 3    Hab2  ~  Hab   1.183981   0.0306792   38.5923        0
## 4    Hab3  ~  Hab   1.127639   0.0304292   37.0578        0
```

10.2.5   Survival analysis

The lifelines package in Python is designed to support survival analysis, with functions to calculate survival estimates, plot survival curves, perform Cox proportional hazard regression and check proportional hazard assumptions. A full tutorial is available at https://lifelines.readthedocs.io/en/latest/index.html.

Here is an example of how to plot Kaplan-Meier survival curves in Python using our Chapter 9 walkthrough example. The survival curves are displayed in Figure 10.1.

```python
import pandas as pd
from lifelines import KaplanMeierFitter
from matplotlib import pyplot as plt

# get data
url = "http://peopleanalytics-regression-book.org/data/job_retention.csv"
job_retention = pd.read_csv(url)

# fit our data to Kaplan-Meier estimates
T = job_retention["month"]
E = job_retention["left"]
kmf = KaplanMeierFitter()
kmf.fit(T, event_observed = E)

# split into high and not high sentiment
highsent = (job_retention["sentiment"] >= 7)


# set up plot
survplot = plt.subplot()

# plot high sentiment survival function
kmf.fit(T[highsent], event_observed = E[highsent],
label = "High Sentiment")

kmf.plot_survival_function(ax = survplot)

# plot not high sentiment survival function
kmf.fit(T[~highsent], event_observed = E[~highsent],
label = "Not High Sentiment")

kmf.plot_survival_function(ax = survplot)

# show survival curves by sentiment category
plt.show()
```

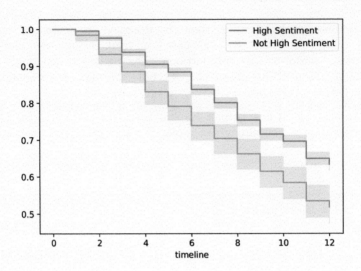

FIGURE 10.1: Survival curves by sentiment category in the job retention data

And here is an example of how to fit a Cox Proportional Hazard model similarly to Section 9.2[1].

```python
from lifelines import CoxPHFitter

# fit Cox PH model to job_retention data
cph = CoxPHFitter()
cph.fit(job_retention, duration_col = 'month', event_col = 'left',
        formula = "gender + field + level + sentiment")


# view results
cph.print_summary()
```

---

[1]I am not aware of any way of running frailty models currently in Python.

```
## <lifelines.CoxPHFitter: fitted with 3770 total observations, 2416 right-censored observations>
##              duration col = 'month'
##                 event col = 'left'
##        baseline estimation = breslow
##     number of observations = 3770
## number of events observed = 1354
##    partial log-likelihood = -10724.52
##          time fit was run = 2021-04-19 09:41:11 UTC
##
## ---
##                           coef  exp(coef)  se(coef)  coef lower 95%  coef upper 95%  exp(coef) lower 95%  exp(coef) upper 95%
## covariate
## gender[T.M]              -0.05       0.96      0.06           -0.16            0.07                 0.85                 1.07
## field[T.Finance]          0.22       1.25      0.07            0.09            0.35                 1.10                 1.43
## field[T.Health]           0.28       1.32      0.13            0.03            0.53                 1.03                 1.70
## field[T.Law]              0.11       1.11      0.15           -0.18            0.39                 0.84                 1.48
## field[T.Public/Government] 0.11      1.12      0.09           -0.06            0.29                 0.94                 1.34
## field[T.Sales/Marketing]  0.09       1.09      0.10           -0.11            0.29                 0.89                 1.33
## level[T.Low]              0.15       1.16      0.09           -0.03            0.32                 0.97                 1.38
## level[T.Medium]           0.18       1.19      0.10           -0.02            0.38                 0.98                 1.46
## sentiment                -0.12       0.89      0.01           -0.14           -0.09                 0.87                 0.91
##
##                              z        p  -log2(p)
## covariate
## gender[T.M]              -0.77     0.44      1.19
## field[T.Finance]          3.34   <0.005     10.24
## field[T.Health]           2.16     0.03      5.02
## field[T.Law]              0.73     0.47      1.10
## field[T.Public/Government] 1.29    0.20      2.35
## field[T.Sales/Marketing]  0.86     0.39      1.36
## level[T.Low]              1.65     0.10      3.32
## level[T.Medium]           1.73     0.08      3.58
## sentiment                -8.41   <0.005     54.49
## ---
## Concordance = 0.58
## Partial AIC = 21467.04
## log-likelihood ratio test = 89.18 on 9 df
## -log2(p) of ll-ratio test = 48.58
```

Proportional Hazard assumptions can be checked using the check_assumptions() method[2].

```
cph.check_assumptions(job_retention, p_value_threshold = 0.05)
```

---

[2]Schoenfeld residual plots can be seen by setting show_plots = True in the parameters.

```
## The ``p_value_threshold`` is set at 0.05. Even under the null hypothesis of no violations, some
## covariates will be below the threshold by chance. This is compounded when there are many covariates.
## Similarly, when there are lots of observations, even minor deviances from the proportional hazard
## assumption will be flagged.
##
## With that in mind, it's best to use a combination of statistical tests and visual tests to determine
## the most serious violations. Produce visual plots using ``check_assumptions(..., show_plots=True)``
## and looking for non-constant lines. See link [A] below for a full example.
##
## <lifelines.StatisticalResult: proportional_hazard_test>
##  null_distribution = chi squared
## degrees_of_freedom = 1
##              model = <lifelines.CoxPHFitter: fitted with 3770 total observations, 2416 right-
censored observations>
##          test_name = proportional_hazard_test
##
## ---
##                                 test_statistic     p  -log2(p)
## field[T.Finance]           km           1.20  0.27      1.88
##                            rank         1.09  0.30      1.76
## field[T.Health]            km           4.27  0.04      4.69
##                            rank         4.10  0.04      4.54
## field[T.Law]               km           1.14  0.29      1.81
##                            rank         0.85  0.36      1.49
## field[T.Public/Government] km           1.92  0.17      2.59
##                            rank         1.87  0.17      2.54
## field[T.Sales/Marketing]   km           2.00  0.16      2.67
##                            rank         2.22  0.14      2.88
## gender[T.M]                km           0.41  0.52      0.94
##                            rank         0.39  0.53      0.91
## level[T.Low]               km           1.53  0.22      2.21
##                            rank         1.52  0.22      2.20
## level[T.Medium]            km           0.09  0.77      0.38
##                            rank         0.13  0.72      0.47
## sentiment                  km           2.78  0.10      3.39
##                            rank         2.32  0.13      2.97
##
##
## 1. Variable 'field[T.Health]' failed the non-proportional test: p-value is 0.0387.
##
##    Advice: with so few unique values (only 2), you can include `strata=['field[T.Health]', ...]` in
## the call in `.fit`. See documentation in link [E] below.
##
## ---
## [A]  https://lifelines.readthedocs.io/en/latest/jupyter_notebooks/
##          Proportional%20hazard%20assumption.html
## [B]  https://lifelines.readthedocs.io/en/latest/jupyter_notebooks/
##          Proportional%20hazard%20assumption.html#Bin-variable-and-stratify-on-it
## [C]  https://lifelines.readthedocs.io/en/latest/jupyter_notebooks/
##          Proportional%20hazard%20assumption.html#Introduce-time-varying-covariates
## [D]  https://lifelines.readthedocs.io/en/latest/jupyter_notebooks/
##          Proportional%20hazard%20assumption.html#Modify-the-functional-form
## [E]  https://lifelines.readthedocs.io/en/latest/jupyter_notebooks/
##          Proportional%20hazard%20assumption.html#Stratification
##
## []
```

10.2.6   Other model variants

Implementation of other model variants featured in earlier chapters becomes thinner in Python. However, of note are the following:

- Ordinal regression is not currently available in the release version of the statsmodels package but is available in the development version. The mord package offers an implementation of ordinal regression for predictive analytics purposes, but for inferential modeling users will need to wait for a release of `statsmodels` that contains ordinal regression methods or for immediate use they will need to install the development version from source.

- Mixed models only currently have an implementation for linear mixed modeling in `statsmodels`. Generalized linear mixed models equivalent to those found in the `lme4` R package are not yet available in Python.

11

Power Analysis to Estimate Required Sample Sizes for Modeling

In the vast majority of situations in people analytics, researchers and analysts have limited control over the size of their samples. The most common situation is, of course, that analyses are run with whatever data can be gleaned and cleaned in the time available. At the same time, as we have seen in all of our previous work, even if a certain difference might exist in real life in the populations being studied, it is by no means certain that a specific analysis on samples from these populations will elucidate that difference. Whether that difference is visible depends on the statistical properties of the samples used. Therefore, researchers and analysts are living in the reality that when they conduct inferential analysis, the usefulness of their work depends to a very large degree on the samples they have available.

This suggests that a conscientious analyst would be well advised to do some up-front work to determine if their samples have a chance of yielding results that are of some inferential value. In a practical context, however, this is only partly true (and that is an important reason why this chapter has been left towards the end of this book). Estimating required sample sizes is an imprecise science. Although the mathematics suggest that in theory it should be precise, in reality we are guessing most of the inputs to the mathematics. In many cases we are so clueless about those inputs that we move into the realms of pure speculation and produce ranges of required sample sizes that are so wide as to be fairly meaningless in practice.

That said, there are situations where conducting *power analysis*—that is, analysis of the required statistical properties of samples in order to have a certain minimum probability of observing a true difference—makes sense. Power analysis is an important element of experimental design. Experiments in people analytics usually take one of two forms:

1. *Prospective experiments* involve running some sort of test or pilot on populations to determine if a certain measure has a hypothesized effect. For example, introducing a certain new employee benefit for a specific subset of the company for a limited period of time, and determining if there was a difference in the impact on employee satisfaction compared to those who did not receive the benefit.

DOI: 10.1201/9781003194156-11

2. *Retrospective experiments* involve the use of historical data to test if a certain measure has a hypothesized effect. This usually occurs opportunistically when it is apparent that a certain measure has occurred in the past and for a limited time, and data can be drawn to test whether or not that measure resulted in the hypothesized effect.

Both prospective and retrospective experiments can involve a lot of work—either in setting up experiments or in extracting data from history. There is a natural question as to whether the chances of success justify the required resources and effort. Before proceeding in these cases, it is sensible to get a point of view on the likely power of the experiment and what level of sample size might be needed in order to establish a meaningful inference. For this reason, power analysis is a common component of research proposals in the medical or social sciences.

Power analysis is a relatively blunt instrument whose primary value is to make sure that substantial effort is not being wasted on foolhardy research. If the analyst already has reasonably available data and wants to test for the effect of a certain phenomenon, the most direct approach is to just go and run the appropriate model assuming that it is relatively straightforward to do so. Power analysis should only be considered if there is clearly some substantial labor involved in the proposed modeling work.

11.1   Errors, effect sizes and statistical power

Before looking at practical ways to conduct power tests on proposed experiments, let's review an example of the logical and mathematical principles behind power testing, so that we understand what the results of power tests mean. Recall from Section 3.3 the logical mechanisms for testing hypotheses of statistical difference. Given data on samples of two groups in a population, the *null hypothesis* H_0 is the hypothesis that a difference does not exist between the groups in the overall population. If the null hypothesis is rejected, we accept the alternative hypothesis H_1 that a difference does exist between the groups in the population.

Recall also that we use the statistical properties of the samples to make inferences about the null and alternative hypotheses based on statistical likelihood. This means that four possible situations can occur when we run hypothesis tests:

1. We fail to reject H_0, and in fact H_1 is false. This is a good outcome.

2. We reject H_0, but in fact H_1 is false. This is known as a *Type I error*.

3. We fail to reject H_0, but in fact H_1 is true. This is known as a *Type II error*.

4. We reject H_0, and in fact H_1 is true. This is a good outcome and one which is most often the motivation for the hypothesis test in the first place.

Statistical power refers to the fourth situation and is the *probability that H_0 is rejected and H_1 is true*. Statistical power depends *at a minimum* on three criteria:

- The significance level α at which the analysis wishes to reject H_0 (see Section 3.3). Usually $\alpha = 0.05$.
- The size n of the sample being used.
- The size of the difference observed in the sample, known as the *effect size*. There are numerous definitions of the effect size that depend on the specific type of power test being conducted.

As an example to illustrate the mathematical relationship between these criteria, let's assume that we run an experiment on a group of employees of size n where we introduce a new benefit and then test their satisfaction levels before and after its introduction. As a statistic of a random variable, we can expect the mean difference in satisfaction to have a normal distribution. Let μ_0 be the mean of the population under the null hypothesis and let μ_1 be the mean of the population under the alternative hypothesis. Now let's assume that in our sample we observe a mean satisfaction of μ^* after the experiment. Recall from Chapter 3 that to meet a statistical significance standard of α, we will need μ^* to be greater than a certain multiple of the standard error $\frac{\sigma}{\sqrt{n}}$ above μ_0 based on the normal distribution. Let's call that multiple z_α. Therefore, we can say that the statistical power of our hypothesis test is:

$$
\begin{aligned}
\text{Power} &= P(\mu^* > \mu_0 + z_\alpha \frac{\sigma}{\sqrt{n}} | \mu = \mu_1) \\
&= P(\frac{\mu^* - \mu_1}{\frac{\sigma}{\sqrt{n}}} > -\frac{\mu_1 - \mu_0}{\frac{\sigma}{\sqrt{n}}} + z_\alpha | \mu = \mu_1) \\
&= 1 - \Phi(-\frac{\mu_1 - \mu_0}{\frac{\sigma}{\sqrt{n}}} + z_\alpha) \\
&= 1 - \Phi(-\frac{\mu_1 - \mu_0}{\sigma}\sqrt{n} + z_\alpha) \\
&= 1 - \Phi(-d\sqrt{n} + z_\alpha)
\end{aligned}
$$

where Φ is the cumulative normal probability distribution function, and $d = \frac{\mu_1 - \mu_0}{\sigma}$ is known as *Cohen's effect size*. Therefore, we can see that power depends on a measure of the observed effect size between our two samples (defined as Cohen's d) the significance level α and the sample size n[1].

The reader may immediately observe that many of these measures are not known at the typical point at which we would wish to do a power analysis. We can assert a minimum level of statistical power that we would wish for— usually this is somewhere between 0.8 and 0.9. We can also assert our α. But at a point of experimental design, we usually do not know the sample size and we do not know what difference would be observed in that sample (the effect size). This implies that we are dealing with a single equation with more than one unknown, and this means that there is no unique solution[2]. Practically speaking, looking at ranges of values will be common in power analysis.

11.2   Power analysis for simple hypothesis tests

Usually we will run power analyses to get a sense of required sample sizes. Given the observations on unknowns in the previous section, we will have to assert certain possible statistical results in order to estimate required sample sizes. Most often, we will need to suggest the observed effect size in order to obtain the minimum sample size for that effect size to return a statistically significant result at a desired level of statistical power.

Using our example from the previous section, let's assume that we would see a 'medium' effect size on our samples. *Cohen's Rule of Thumb* for d states that $d = 0.2$ is a small effect size, $d = 0.5$ a medium effect size and $d = 0.8$ a large effect size. We can use the `wp.t()` function from the WebPower package in R to do a power analysis on a paired two-sample t-test and return a minimum required sample size. We can assume $d = 0.5$ and that we require a power of 0.8—that is, we want an 80% probability that the test will return an accurate rejection of the null hypothesis.

```
library(WebPower)

# get minimum n for power of 0.8
(n_test <- WebPower::wp.t(d = 0.5, p = 0.8, type = "paired"))
```

---

[1] We will also need to know the expected distribution of the statistics that we are analyzing in order to determine the power probability.

[2] In reality there are more unknowns that this math would imply, due to the imperfection of what we are trying to measure. For example measurement error and reliability will often be an unmeasurable unknown. For this reason you will often need a larger sample size than that indicated by power tests.

```
## Paired t-test
##
##              n    d alpha power
##      33.36713 0.5  0.05   0.8
##
## NOTE: n is number of *pairs*
## URL: http://psychstat.org/ttest
```

This tells us that we need an absolute minimum of 34 individuals in our sample for an effect size of 0.5 to return a significant difference at an alpha of 0.05 with 80% probability. Alternatively we can test the power of a specific proposed sample size.

```
# get power for n of 40
(p_test <- WebPower::wp.t(n1 = 40, d = 0.5, type = "paired"))
```

```
## Paired t-test
##
##       n   d alpha      power
##      40 0.5  0.05 0.8693981
##
## NOTE: n is number of *pairs*
## URL: http://psychstat.org/ttest
```

This tells us that a minimum sample size of 40 would result in a power of 0.87. A similar process can be used to plot the dependence between power and sample size under various conditions as in Figure 11.1. This is known as a *power curve*.

```
# test a range of sample sizes
sample_sizes <- 20:100
power <- WebPower::wp.t(n1 = sample_sizes, d = 0.5, type = "paired")

plot(power)
```

We can see a 'sweet spot' of approximately 40–60 minimum required participants, and a diminishing return on statistical power over and above this. Similarly we can plot a proposed minimum sample size against a range of effect sizes as in Figure 11.2.

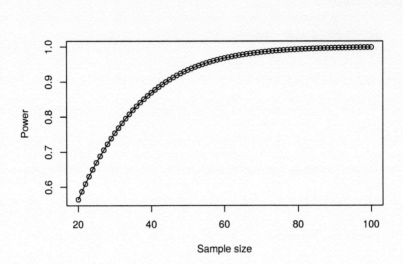

FIGURE 11.1: Plot of power against sample size for a paired t-test

```
# test a range of effect sizes
effect_sizes <- 2:8/10
samples <- WebPower::wp.t(n1 = rep(40, 7),
                         d = effect_sizes,
                         type = "paired")
plot(samples$d, samples$power, type = "b",
     xlab = "Effect size", ylab = "Power")
```

Similar power test variants exist for other common simple hypothesis tests. Let's assume that we want to institute a screening test in a recruiting process, and we want to validate this test by running it on a random set of employees with the aim of proving that the test score has a significant non-zero correlation with job performance. If we assume that we will see a moderate correlation of $r = 0.3$ in our sample[3], we can use the wp.correlation() function in WebPower to do a power analysis, resulting in Figure 11.3.

```
sample_sizes <- 50:150
correl_powers <- WebPower::wp.correlation(n = sample_sizes, r = 0.3)
plot(correl_powers)
```

---

[3]Cohen's rule of thumb for correlation coefficients is Weak: 0.1, Moderate: 0.3 and Strong: 0.5.

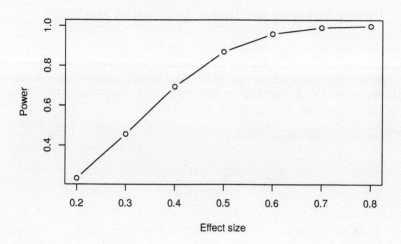

FIGURE 11.2: Plot of power against effect size for a paired t-test

Figure 11.3 informs us that we will likely want to be hitting at least 100 employees in our study to have any reasonable chance of establishing possible validity for our screening test.

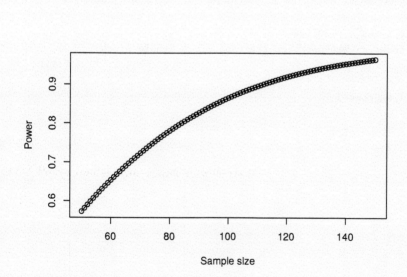

FIGURE 11.3: Plot of power against sample size for a correlation test

11.3   Power analysis for linear regression models

In power tests of linear regression models, the effect size is a statistic of the difference in model fit between the two models being compared. Most commonly this will be a comparison of a 'full' fitted model involving specific input variables compared to a 'reduced' model with fewer input variables (often a random variance model with no input variables).

The f^2 statistic is defined as follows:

$$f^2 = \frac{R^2_{\text{full}} - R^2_{\text{reduced}}}{1 - R^2_{\text{full}}}$$

where the formula refers to the R^2 fit statistics for the two models being compared. As an example, imagine we already know that GPA in college has a significant relationship with job performance, and we wish to determine if our proposed screening test had incremental validity on top of knowing college GPA. We might run two linear regression models, one relating job performance to GPA, and another relating job performance to *both* GPA and screening test score. Assuming we would observe a relatively small effect size for our screening test, we assume $f^2 = 0.05$[4], we can plot sample size against power in determining whether the two models are significantly different. We will also need to define the number of predictors in the full model (p1 = 2) and the reduced model (p2 = 1). The plot is shown in Figure 11.4.

```
sample_sizes <- 100:300
f_sq_power <- WebPower::wp.regression(n = sample_sizes,
                         p1 = 2, p2 = 1, f2 = 0.05)

plot(f_sq_power)
```

---

[4]Cohen's rule of thumb for f^2 effect sizes is Small: 0.02, Medium: 0.15, Large: 0.35.

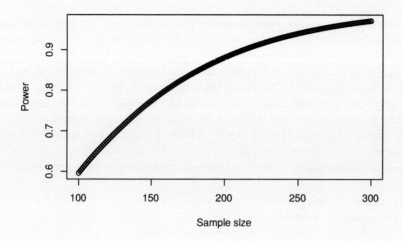

FIGURE 11.4: Plot of power against sample size for a small effect of a second input variable in a linear regression model

11.4   Power analysis for log-likelihood regression models

In Chapter 5, we reviewed how measures of fit for log-likelihood models are still the subject of some debate. Given this, it is unsurprising that measures of effect size for log-likelihood models are not well established. The most well-developed current method appeared in Demidenko (2007), and works when we want to do a power test on a single input variable x using the Wald test on the significance of model coefficients (see Section 7.3.2 for a reminder of the Wald test).

In this method, the statistical power of a significance test on the input variable x is determined using multiple inputs as follows:

1. The likelihood of a positive outcome when $x = 0$ is used to determine the intercept (`p0` in the code below).
2. The likelihood of a positive outcome when $x = 1$ is then used to determine the regression coefficient for x (`p1` in the code below).
3. A distribution for x is inputted (`family` below) and the parameters of that distribution are also entered (`parameter` below). For example, if the distribution is assumed to be normal then the mean and standard deviation would be entered as parameters.

4. This information is fed into the Wald test, and the power for specific sample sizes is calculated.

For example, let's assume that we wanted to determine if our new screening test had a significant effect on promotion likelihood by running an experiment on employees who were being considered for promotion. We assume that our screening test is scored on a percentile scale and has a mean of 53 and a standard deviation of 21. We know that approximately 50% of those being considered for promotion will be promoted, and we believe that the screening test may have a small effect whereby those who score zero would still have a 40% chance of promotion and every additional point scored would increase this chance by 0.2 percentage points. We run the `wp.logistic()` function in WebPower to plot a power curve for various sample sizes as in Figure 11.5.

```
sample_sizes <- 50:2000
logistic_power <- WebPower::wp.logistic(n = sample_sizes,
                                        p0 = 0.4, p1 = 0.402,
                                        family = "normal",
                                        parameter = c(53, 21))

plot(logistic_power)
```

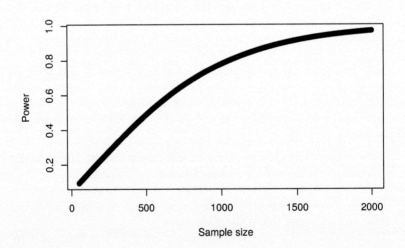

FIGURE 11.5: Plot of power against sample size for a single input variable in logistic regression

This test suggests that we would need over 1000 individuals in our experiment in order to have at least an 80% chance of establishing the statistical significance of a true relationship between screening test score and promotion likelihood.

11.5 Power analysis for hierarchical regression models

Power tests for explicit hierarchical models usually originate from the context of the design of clinical trials, which not only concern themselves with the entire sample size of a study but also need to determine the split of that sample between treatment and control. It is rare that power analysis would need to be conducted for hierarchical models in people analytics but the technology is available in the WebPower package to explore this.

Cluster randomized trials are trials where it is not possible to allocate individuals randomly to treatment or control groups and where entire clusters have been allocated at random instead. This creates substantial additional complexity in understanding statistical power and required sample sizes. The wp.crt2arm() function in WebPower supports power analysis on 2-arm trials (treatment and control), and the wp.crt3arm() function supports power analysis on 3-arm trials (Two different treatments and a control).

Multisite randomized trials are trials where individuals are assigned to treatment or control groups at random, but where these individuals also belong to different clusters which are important in modeling—for example, they may be members of clinical groups based on pre-existing conditions, or they may be being treated in different hospitals or outpatient facilities. Again, this makes for a substantially more complex calculation of statistical power. The wp.mrt2arm() and wp.mrt3arm() functions offer support for this.

Power tests are also available for structural equation models. This involves comparing a more 'complete' structural model to a 'subset' model where some of the coefficients from the more 'complete' model are set to zero. Such power tests can be valuable when structural models have been applied previously on responses to survey instruments and there is an intention to test alternative models in the future. They can provide information on required future survey participation and response rates in order to establish whether the improved fit can be established for the alternative models.

There are two approaches to power tests for structural equation models, using a chi square test and a root mean squared error (RMSEA) approach. Both of these methods take a substantial number of input parameters, consistent with the complexity of structural equation model parameters and the various alternatives for measuring fit of these models. The chi square test approach is implemented by the `wp.sem.chisq()` function, and the RMSEA approach is implemented by the `wp.sem.rmsea()` function in `WebPower`.

11.6   Power analysis using Python

A limited set of resources for doing power analysis is available in the `stats.power` module of the `statsmodels` package. As an example, here is how we would conduct the power analysis for a paired *t*-test as in Section 11.2 above.

```
import math
from statsmodels.stats.power import TTestPower

power = TTestPower()
n_test = power.solve_power(effect_size = 0.5,
                           power = 0.8,
                           alpha = 0.05)
print(math.ceil(n_test))
```

```
## 34
```

And a power curve can be constructed as in Figure 11.6.

```
import matplotlib.pyplot as plt
import numpy as np

fig = plt.figure()
fig = TTestPower().plot_power(dep_var = 'nobs',
                              nobs = np.arange(20, 100),
                              effect_size = np.array([0.5]),
                              alpha = 0.05)
plt.show()
```

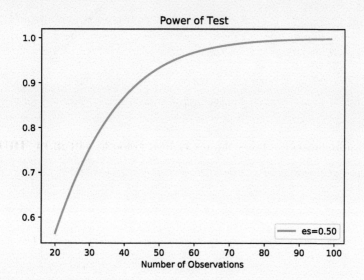

FIGURE 11.6: Plot of power against sample size for a paired t-test

12

Further Exercises for Practice

This final chapter contains a set of scenarios and exercises that will allow you to put into further practice some of the techniques you have learned in this book, and are supplementary to the exercises provided at the end of each of the earlier chapters. All the exercises are based on data that is available in the peopleanalyticsdata package in R, or alternatively can be downloaded from the internet. While the scenarios are fictitious, they are intended to represent typical questions and situations that arise when doing statistical modeling in people analytics.

As you work through these scenarios, I encourage you to document your work using either an R Markdown document or a Jupyter Notebook if you prefer. This will help you keep a record of your method, approach and code in case you need to put it into practice again in the future. It will also make it easy for you to share your work with others (for example by putting it in a Github repository), which will allow you to collaborate, discuss and open your work to critique. If you are starting out on your analytics journey, exposing your work to others is one of the best ways to learn. If you are more experienced, then there are others that will undoubtedly benefit from seeing how you went about solving these problems.

12.1 Analyzing graduate salaries

Graduate salary levels are important economic indicators of the value of tertiary education. They can provide important insight about the value of education to employers or to the economy as a whole. When studied in detail they can highlight which particular disciplines have higher or lower levels of demand for graduates, and they can be important factors in determining what subjects or majors students choose to specialize in.

Government agencies and educational institutions will analyze graduate salaries regularly to help critique or validate policy or strategy. Employers will also regularly study publicly available graduate salary information to help

DOI: 10.1201/9781003194156-12

them benchmark their compensation and benefits against the external market
for graduates.

12.1.1 The graduates data set

The graduates data set contains information on graduates currently in the
United States across 173 specific subject majors grouped into 16 disciplines of
study. This data set is sourced from the *FiveThirtyEight* data repository[1].

Load the graduates data set via the peopleanalyticsdata package or download
it from the internet[2]. The fields in the graduate data set are as follows:

- Major is the specific subject major.
- Discipline is the broad subject discipline.
- Total is the number of graduates of working age in the US.
- Unemployment_rate is the proportion of graduates currently unemployed.
- Median_salary is the current median salary of those employed in US dollars.

12.1.2 Discussion questions

1. What kind of outcome is the Median_salary column?
2. Which of the variables in the data set would you be interested in
   using to explain the Median_salary outcome? Why?
3. Are there any transformations you would consider on any of the
   input variables to help with interpreting the model?
4. What type of model would you use to try to explain the Me-
   dian_salary outcome using these variables?
5. Describe the data type of each of the input variables. How would
   you interpret the coefficients of the model for each of these data
   types?

12.1.3 Data exercises

1. Perform an exploratory data analysis on the data set. Do you see
   any interesting patterns?
2. Conduct appropriate hypothesis tests on any of the input variables
   that interest you to determine if they relate to statistically significant differences in the Median_salary outcome.
3. If you wish to, perform transformations on the data to help with

---

[1] https://github.com/fivethirtyeight/data
[2] http://peopleanalytics-regression-book.org/data/graduates.csv

the interpretation of model results. For example, are the numerical scales intuitive for interpretation?

4. Run an appropriate multivariate model to determine which input variables have a significant effect on median graduate salary.

5. Articulate the results of your model, including an estimate of the effect of the input variables and the overall fit and goodness-of-fit of the model.

12.2 Analyzing a recruiting process

Organizations are often very interested in analyzing data from recruiting processes, usually with a couple of goals in mind. Firstly, there is an interest in whether the process is efficient and effective. Secondly, there is an interest in whether the process is fair to different groups and a foundation for the recruiting of a sufficiently diverse set of employees.

The efficiency and effectiveness of a recruiting process can depend heavily on how it is organized, what methods are used and whether those methods are helpful in determining hiring decisions. Statistics from individual elements of the process such as interviews or assessments are often studied. Models can be built to determine which elements influence the decision to hire or not to hire. In an ideal world, some sort of future job performance outcome would be particularly useful in studying the efficiency and effectiveness of a recruiting process, but often this is a very difficult thing to do, especially if the process is very selective. Hiring only a small proportion of applicants usually results in a statistical phenomenon known as *range restriction*, where the statistics of those hired fall in a very narrow range that makes useful analysis extremely challenging. For this reason, many organizations focus primarily on the final hiring decision as an outcome of interest.

Understanding fairness in a recruiting process usually involves studying how the statistics of that process differ between subgroups of interest and whether any of the differences are significant enough to infer potential bias. Understanding whether these differences are attributable to a particular element of the process, whether it be a test or the rating behaviors of interviewers, is also important in determining whether specific action can be taken to remedy the situation.

12.2.1 The recruiting data set

The recruiting data set contains information on 966 applicants who went through the final stages of a recruiting process for graduate positions at a large international financial services company. The recruiting process operates as follows:

1. Applications are screened according to a number of criteria, including their SAT scores and their undergraduate GPA as well as an online aptitude test they are requested to take and numerous other judgments made by the individuals screening the applications.
2. Applicants who pass the screening stage are invited for three interviews, two of them with line managers and a third with a human resources professional. Different line managers or HR professionals conduct the interviews on different interview days. Each interviewer independently gives an applicant a score on a Likert scale of 1–5 indicating increasing positive sentiment.
3. Interviewers and human resources professionals gather to discuss each case and make a final decision on whether to hire or not to hire. All the information used in screening and evaluating applicants is made available to decision makers during this discussion.

Load the recruiting data set via the peopleanalyticsdata package or download it from the internet[3]. The fields in the recruiting data set are as follows:

- gender is the gender of the applicant.
- sat is the SAT score of the applicant.
- gpa is the Undergraduate GPA of the applicant.
- apttest is the score on the aptitude test given to the applicant.
- int1 is the rating of the first line manager interviewer.
- int2 is the rating of the second line manager interviewer.
- int3 is the rating of the human resources interviewer.
- hired is a binary indicator of whether a decision was made to hire the applicant.

12.2.2 Discussion questions

1. Considering the way the recruiting process works, what kinds of inferential analysis or modeling would you be interested in applying to help understand its efficiency and effectiveness?
2. What kind of model is most appropriate for explaining the hired outcome?

---

[3]http://peopleanalytics-regression-book.org/data/recruiting.csv

3. One of your stakeholders is suggesting that the aptitude test is a waste of time and that the information it provides can already be gleaned from the applicants' SAT scores and GPA. What kind of statistical analysis or model would help you confirm or reject this?

4. Do you think that collinearity might pose a risk in this data? If so, what variables would concern you?

5. What kind of hypothesis test would you use to determine if the hiring outcome may be different by gender?

6. What kind of hypothesis test would you use to determine if the aptitude test score may be different by gender?

7. How would you go about determining if any gender difference in the hiring outcome can be attributed to a specific part of the process?

12.2.3 Data exercises

1. Perform an exploratory data analysis on the `recruiting` data set. Be sure to convert data to the best type for your purposes.

2. Develop a model to test how the aptitude test results are explained by an applicant's SAT and GPA. What can you conclude from this? Have you considered possible collinearity in this model?

3. Develop a model to explain how all the inputs in the hiring decision (interview ratings, aptitude test, SAT and GPA) influence the hiring decision. Reduce this to the most parsimonious model you are comfortable with. What can you conclude from this model about the role of the different elements of the recruiting process in the final hiring decision?

4. Test whether there is a statistically significant difference in the hiring outcome for males versus females.

5. By adding gender into your model from Data Exercise 3, determine what element or elements of the recruiting process may be related to any differences in gender in the hiring outcome.

12.3 Analyzing the drivers of performance ratings

In many organizations and for many job types, promotion and performance ratings are the primary indicators of the success of employees. However, promotion is not always available to employees and can be very dependent on role and timing. Since performance ratings are usually generated on a regular basis, it is usually these that garner the most attention in the analysis of success.

However, performance ratings are not perfect indicators of reality. They are usually the result of some judgment from one or more evaluators. Part of that judgment will be informed by data and part will be informed by contextual considerations or personal preferences outside of the data. Therefore it is frequently of interest to analyze performance ratings as the outcome of a decision making process. Such analysis can inform us as to what parts of the evaluation process are operating as intended and what parts are not. Multivariate models around performance can help us understand the degree to which the evaluation process is data driven, the degree to which unfairness might exist in some evaluation decisions and what might be the source of that unfairness.

12.3.1   The `employee_performance` data set

The `employee_performance` data set contains data on the most recent performance evaluations of 366 salespeople in a technology company. Each employee is evaluated by their manager, who considers certain performance indicators together with their own judgment and awards a performance rating from 1 to 3, where 1 means 'Needs improvement,' 2 means 'Performing well' and 3 means 'Outstanding.'

Load the `employee_performance` data set via the `peopleanalyticsdata` package or download it from the internet[4]. The fields in the `employee_performance` data set are as follows:

- `sales`: The sales in millions of dollars made by the salesperson during the evaluation period
- `new_customers`: The number of new customers acquired by the salesperson during the evaluation period
- `region`: The region that the salesperson operates in: North, South, East or West
- `gender`: The gender of the salesperson
- `rating`: The performance rating awarded by the manager

12.3.2   Discussion questions

1. What hypothesis test should be used to determine if there is a significant relationship between sales and performance rating?
2. What hypothesis test should be used to determine if there is a similar distribution of performance ratings between the four regions?
3. What type of an outcome is the performance rating? What kind of model is appropriate for explaining what influences performance ratings?

---

[4]http://peopleanalytics-regression-book.org/data/employee_performance.csv

4. Which input variables would you want to be significant and which would you want to be insignificant in your model in order to support an argument that the evaluation process is fair and relevant to the job?
5. What assumptions would you need to check after you have run your model to have confidence that you can trust your inferences?

12.3.3 Data exercises

1. Run separate and appropriate hypothesis tests on each of the input variables to determine if they have a significant effect on the performance outcome.
2. Prepare your data for running an appropriate multivariate model to explain the performance outcome. Be sure to convert to appropriate data types.
3. Run an appropriate multivariate model to explain the performance outcome. Report on which variables are significant and estimate the effects of the significant variables.
4. Determine and comment on the overall fit and goodness-of-fit of your model. Use this to make a comment on how 'data driven' you believe the evaluation process to be.
5. Perform appropriate checks on the assumptions of your model. What approach might you take if any of these tests fail?

12.4 Analyzing promotion differences between groups

As mentioned in the previous section, promotion is a more challenging outcome to analyze because it can happen at different times for different people or groups. Nevertheless, over certain time periods—usually several years—organizations will be interested in understanding what affects the likelihood of promotion among their employees.

It is particularly interesting to compare subgroups of employees that have common characteristics to see if there is a difference in their likelihood of promotion over a specified time period. This is highly analogous to the study of retention or attrition in that promotion can be considered a singular event that can happen to different individuals at different points in time. Like retention or attrition, we are not only interested in whether this event occurred as at the end of a period of time, we are also interested in when it occurred throughout the period.

12.4.1   The promotion data set

The promotion data set contains data on 1134 individuals who joined a re-
tailer in an entry-level job, and tracks them for up to eight subsequent years
post joining. For each individual the data records whether or not they were
promoted, and if so in which year the promotion occurred, where the date
of their joining is Year 0. Once promotion occurs, an individual is no longer
tracked. If an individual was not promoted, then the year in which the last
record occurred is captured.

Load the promotion data set via the peopleanalyticsdata package or download
it from the internet[5]. The fields in the data are as follows:

- diverse indicates whether or not the individual is a member of one of the
  organization's diversity programs.
- flexible indicates whether the individual worked on a part-time program
  for at least 6 months.
- store indicates if the individual joined the company in a retail store position.
- promoted indicates whether the individual was promoted.
- year indicates the year in which the last record was made of the individual.

12.4.2   Discussion questions

1. What type of analysis is most appropriate to understand if there
   is a difference in promotion likelihood for employees who are on
   flexible hours, in diversity programs or who work in-store?
2. What type of illustration would you use to show whether each of
   these three factors individually have an effect on promotion likelihood?
3. What type of model would be most appropriate to determine the
   combined effects of all of these factors on promotion likelihood?
4. How would you go about determining whether any differences in
   promotion likelihood in a given group (such as the flexible working
   group) is mediated by membership of another group?
5. What assumptions would you need to check to validate that your
   analysis is trustworthy?

12.4.3   Data exercises

1. Run an exploratory data analysis to understand any general patterns of interest in the data.

---

[5]http://peopleanalytics-regression-book.org/data/promotion.csv

2. Perform an analysis and generate an appropriate graph to illustrate the impact of flexible working on the likelihood of promotion. Determine if there is a statistically significant effect.

3. Repeat this analysis to determine the impact of diversity program membership.

4. Run an appropriate multivariate model to determine how all three variables of flexible working, diversity and in-store working affect the likelihood of promotion. Remember to check the assumptions of your model.

5. How would you explain your conclusions? Are there any corrective actions that this analysis might point to?

12.5 Analyzing feedback on learning programs

Assessing the effectiveness of learning programs remains one of the most challenging problems in people analytics. As with many challenging analytics problems, measurement is the key issue. It is exceptionally difficult to track and measure the impact of learning on the future day-to-day success of the individuals who have had access to it. While it makes sense to try to understand the influence of learning on important outcomes like employment, promotion or attrition, these outcomes can often be too distant, too generic in nature and too influenced by context and other factors to expect specific learning participation to show any meaningful influence on them.

Because of the challenges in objectively measuring the impact of learning, analysts often rely on the reaction and feedback of learning participants as an important measure of the success of learning programs. If the content of the program is known to be important to future work and related to future success, and if the participants report that the program was effective for them, then this can create a compelling argument for the success of the program.

12.5.1 The learning data set

The learning data set contains 4974 instances of feedback from 326 different participants in a range of learning programs offered by an executive education provider. Each row of data represents the feedback of a specific participant on a specific program that they participated in. Participants were not required to respond to all feedback questions and any question where no response was given is indicated as NA.

Load the learning data set via the peopleanalyticsdata package or download it from the internet[6]. The fields in the learning data set are as follows:

- idcode is the unique ID of the participant.
- rec is a binary value indicating whether the participant would recommend the program to others.
- rel is a rating from the participant on the relevance of the program to their work, where 1 is Very Low and 5 is Very High.
- fun is a rating on how enjoyable and fun the participant found the program, where 1 is Very Low and 5 is Very High
- clar is a rating from the participant on the clarity of the content and teaching in the program, where 1 is Very Low and 5 is Very High.
- home is a rating from the participant on the quality of the homework or project work in the program, where 1 is Very Low and 5 is Very High.
- class is a rating from the participant on the quality of the overall class who attended the program, where 1 is Very Low and 5 is Very High.
- fac is a rating from the participant on the quality of the program faculty and instructors, where 1 is Very Low and 5 is Very High.

12.5.2 Discussion questions

1. What kind of outcome is the rec column and which kind of model best suits this outcome type?
2. Describe the nature of the hierarchy in this data set.
3. Describe what question we would be answering if we ignored the hierarchy in modeling what influences rec.
4. Describe what question we would be answering if we considered the hierarchy in modeling what influences rec.
5. What kind of model would you use to explicitly consider this hierarchy in modeling what influences rec? What kinds of parameters would you experiment with in running this model?
6. Describe what you would expect to see in the output of a model that considered the hierarchy in modeling what influences rec.

12.5.3 Data exercises

1. Prepare and run a model to determine which elements of feedback influence whether or not the program will be recommended to others.
2. Prepare and run a separate model to determine which elements of feedback *influence a participant* in deciding if they would recommend a program to others.

---

[6]http://peopleanalytics-regression-book.org/data/learning.csv

3. Experiment with the model from Data Exercise 2 by adjusting which parameters you model at the participant level.

4. Describe the different outputs of your models from Data Exercises 2 and 3 and how to interpret them.

5. Compare the outputs of your models from Data Exercises 2 and 3 to those from Data Exercise 1. How might your conclusions differ between the two modeling approaches?

References

Agresti, Alan. 2007. *An Introduction to Categorical Data Analysis.*

———. 2010. *Analysis of Ordinal Categorical Data.*

Bartholomew, David J., Martin Knott, and Irini Moustaki. 2011. *Latent Variable Models and Factor Analysis: A Unified Approach.*

Bhattacharya, P. K., and Prabir Burman. 2016. *Theory and Methods of Statistics.*

Collett, David. 2015. *Modelling Survival Data in Medical Research.*

Demidenko, Eugene. 2007. "Sample Size Determination for Logistic Regression Revisited." *Statistics in Medicine.*

Fagerland, Morten W., and David W. Hosmer. 2017. "How to Test for Goodness of Fit in Ordinal Logistic Regression Models." *The Stata Journal.*

Fagerland, Morten W., David W. Hosmer, and Anna M. Bofin. 2008. "Multinomial Goodness-of-fit Tests for Logistic Regression Models." *Statistics in Medicine.*

Hosmer, David W., Stanley Lemeshow, and Rodney X. Sturdivant. 2013. *Applied Logistic Regression.*

Jiang, Jiming. 2007. *Linear and Generalized Linear Mixed Models and Their Applications.*

Menard, Scott. 2010. *Logistic Regression: From Introductory to Advanced Concepts and Applications.*

Montgomery, Douglas C., Elizabeth A. Peck, and G. Geoffrey Vining. 2012. *Introduction to Linear Regression Analysis.*

Rao, C. Radhakrishna, Shalabh, Helge Toutenburg, and Christian Heumann. 2008. *The Multiple Linear Regression Model and Its Extensions.*

Senn, Stephen. 2011. "Francis Galton and Regression to the Mean." *Significance.*

Skrondal, Anders, and Sophia Rabe-Hesketh. 2004. *Generalized Latent Variable Modeling: Multilevel, Longitudinal, and Structural Equation Models.*

Venables, W. N., and B. D. Ripley. 2002. *Modern Applied Statistics with s.*

DOI: 10.1201/9781003194156-12

Wickham, Hadley. 2016. *ggplot2: Elegant Graphics for Data Analysis.* https:
   //ggplot2-book.org/.

Wickham, Hadley, and Garrett Grolemund. 2016. *R for Data Science.* https:
   //r4ds.had.co.nz/.

Xie, Yihui, Christophe Dervieux, and Emily Riederer. 2020. *R Markdown
   Cookbook.* https://bookdown.org/yihui/rmarkdown-cookbook/.

Glossary

alpha In hypothesis testing, the standard for rejection of the null hypothesis (usually 0.05). The p-value of the hypothesis test needs to be less than alpha to reject the null hypothesis.

alternative hypothesis In hypothesis testing, the negation of the null hypothesis. Usually this is the hypothesis that a given statement about a population is true.

binomial logistic regression A regression technique that models the probability of a binary or dichotomous event using a logistic function.

coefficient An estimated parameter of a model which can be used to explain the influence of an input variable on the outcome variable.

collinearity A situation where two input variables are highly correlated with each other. This can affect the accuracy of inferences from models.

confidence interval In statistical estimation, the range of values that is likely to contain the true value of a parameter to a specified level of certainty—most commonly 95%.

correlation A normalized version of covariance that ranges from -1 to 1. It is a more intuitive measure of how one variable changes as another changes.

covariance A measure of the extent to which one variable changes as another changes.

Cox proportional hazard A regression technique used to model singular events that occur at different times, used frequently in survival analysis.

dummy variable A way of expressing the value of a categorical variable in binary form to allow that variable to be used as an input to a model. A variable that takes a set of k categorical values is converted to k dummy binary variables.

effect size A measure of the strength of the relationship between two variables in a sample, used in calculations of statistical power. Numerous measures of effect size exist for different types of analysis. Cohen's d is the most

DOI: 10.1201/9781003194156-12

well known and is a normalized measure of the size of the difference between two sample means.

factor analysis A technique used to confirm or refine a set of latent variables that are proposed to explain a larger number of measured variables.

fit The extent to which the variance of the outcome variable is explained by a model. Also used as a verb to describe the process of calculating the optimal parameters of a model which reduce the error to a minimum.

generalized linear model (GLM) A generalization of linear regression to allow for modeling of outcome variables that are not normally distributed. Logistic regression models which model binary or categorical outcome variables are among the most common GLMs.

goodness-of-fit In linear regression, a test of the null hypothesis that a fitted model has no better explanatory power than a null or random model. In logistic regression, a test of the null hypothesis that a fitted model has a better explanatory power than a null or random model.

hypothesis test A statistical technique that determines the likelihood of truth of a statement about a population based on the statistical properties of a sample of that population.

inferential model A model whose primary objective is to explain the relationships between input variables and an outcome variable.

input variable A variable which is to be used in a model to try to explain the outcome variable. Also known as a covariate or an independent variable.

linear regression A regression technique that models a linear relationship between a set of input variables and a *continuous* outcome variable.

logistic regression A set of regression techniques that model the probabilities of at least two distinct outcomes using a logistic function to approximate a probability distribution.

mean The average value of a sample of data.

mixed model A regression technique that accounts for hierarchical groupings in data by modeling fixed effects within groups and random effects between groups. Also known as a hierarchical or multilevel model.

multicollinearity A situation where more than two input variables have a strong linear relationship with each other. This can affect the accuracy of inferences from models.

multinomial logistic regression A regression technique that models the probability of a number of distinct nominal outcome events relative to a reference outcome event.

normal distribution The theoretical distribution of a random variable in an unbounded bell-shaped curve.

null hypothesis In hypothesis testing, the assumption that a statement about a population is not true. The null hypothesis needs to be rejected to prove statistical significance.

odds The ratio of the probability of an event occurring and the probability of it not occurring. A key metric in interpreting logistic regression models.

odds ratio In logistic regression, the multiple of the odds of the outcome brought about by a unit change in an input variable, assuming no change in the other input variables. It is the exponent of the coefficient of that input variable.

ordinal logistic regression A set of regression techniques that model an ordered category outcome such as a Likert scale. The most commonly used technique is proportional odds logistic regression.

outcome variable The variable to be explained in a model. Also known as a dependent variable, target variable or response variable.

parametric model A model for which the outcome variable can be explained or predicted by the input variables without any need for additional information.

Pseudo-R-squared A variety of metrics used to measure the fit of a logistic regression model, intended to be analogous to the R-squared in linear regression.

p-value In hypothesis testing, the maximum probability of the null hypothesis based on the statistical properties of a sample. p-values need to be lower than a certain standard (known as alpha and usually 0.05) to reject the null hypothesis.

people analytics The study of the behaviors and characteristics of people or groups in relation to important business, organizational or institutional outcomes.

predictive model A model whose primary objective is to accurately predict the outcome variable from new observations of the input variables.

random variable A variable which is independent and identically distributed. The value of an observation of a random variable is completely
independent of the value of other observations.

R-squared A metric used to measure the fit of a linear regression model,
with a value ranging from 0 to 1.

regression A set of statistical techniques used to estimate the relationship
between a set of input variables and an outcome variable.

residual The error or difference between the true value of the outcome variable and that estimated by a model for a given observation.

standard deviation The square root of the variance for a sample of data.
It gives a more intuitive measure of how the data varies around its mean
relative to its inherent scale.

standard error A standard deviation of a sampled statistic of a random variable. For example, the standard error of the mean is the standard deviation
of the sampled means of the variable.

statistical power The probability that a null hypothesis is accurately rejected, used in considering the minimal sample sizes needed in experiments.

stratified models Where the outcome variable is categorical with more than
two categories, "stratified models" refers to a set of binomial models that
model membership of each category.

structural equation model A regression technique that proposes a smaller
number of latent variables to explain the measured input variables and models the relationship between the latent variables and the outcome variable.
Particularly useful in the analysis of large survey instruments.

survival analysis A technique used to model the likelihood of singular
events that occur at different times, so named because of its use in epidemiology to study death and disease events.

variance A measure of the extent to which a sample of data varies around
its mean.

Index

AIC (Akaike Information Criterion), 120

Akaike Information Criterion, *see* AIC

binomial logistic regression, 101–123
    coefficients, 110, 114
    goodness-of-fit tests, 119
    multiple, 113
    predictions, 122
    simple, 106
Brant-Wald test, 157

Cohen's effect size, 223
collinearity, 90, 122
confidence interval (statistics), 49
correlation (statistics), 44
    point-biserial correlation, 45
    rank-biserial correlation, 46
covariance (statistics)
    population covariance, 44
    sample covariance, 43
Cox proportional hazard, 188, 193
    proportional hazard assumption, 196

data sets
    charity_donation, 63, 124
    employee_performance, 240
    employee_survey, 186
    graduates, 236
    health_insurance, 129, 141
    job_retention, 188, 201
    learning, 243
    managers, 160
    politics_survey, 172
    promotion, 242
    recruiting, 238

    salespeople, 39, 104, 204
    soccer, 146, 206
    sociological_data, 97
    speed_dating, 165, 185
    ugtests, 38, 67
descriptive statistics, 40–46
dummy variables
    in linear regression, 84
    in logistic regression, 113

evidence-based practice, 3
experiments, 221

frailty models, 197

generalized linear model, *see* GLM
GLM (generalized linear model), 101

hierarchical data, 163
hypothesis testing, 49–61
    alpha, 50
    alternative hypothesis, 50
    chi-square test, 57
    logic and intuition, 49
    null hypothesis, 49
    one and two tailed tests, 52
    p-hacking, 54
    p-value, 50, 52
    statistically significant, 54
    test of non-zero correlation, 54

inferential model, 2, 3, 5
input variables, 5
    categorical, 84, 113

Kaplan-Meier estimates, 188, 190
Kendall's tau, 45

linear regression, 65–96

DOI: 10.1201/9781003194156-12

coefficients, 70, 76, 86
data sparseness, 83
F-statistic, 80
fitting, 74
goodness-of-fit, 80
homoscedacity of residuals, 88
interaction terms, 93
linearity and additivity
    assumption, 86
multiple linear regression, 76
normal distribution of residuals,
    89
ordinary least squares, 65
origins, 65
predictions, 81
quadratic and polynomial
    extensions, 96
R-squared, 74
residuals, 72
selecting input variables, 83
simple linear regression, 69
log odds, 109
logistic function
as a model for probability, 107
origins, 102

mean (statistics), 40
mixed models, 164–170
fixed effects, 164
random effects, 165
multicollinearity, 91, 122
multilevel models, *see* mixed models
multinomial logistic regression,
    127–140
changing reference, 137
coefficients, 135
goodness-of-fit, 140
Independence of Irrelevant
    Alternatives (IIA)
    assumption, 128
model simplification, 138
reference category, 127, 134
relative risk, 134

normal distribution (statistics), 48

odds, 109, 110
odds ratio, 110
ordinal logistic regression, *see*
    proportional odds logistic
    regression
outcome variable, 5
binary, 103
continuous, 66
dichotomous, 103
nominal categories, 128
ordinal, 143

parametric model, 4
parsimony, 120
Pearson residuals, 122
Pearson's correlation, 44
power analysis, 221–232
f-squared, 228
logic and principles, 222
predictive model, 2
proportional odds logistic regression,
    143–159
alternatives to, 158
coefficients, 151
goodness-of-fit, 154
predictions, 153
proportional odds assumption,
    145, 155
Pseudo-R-squared
in binomial logistic regression,
    118
in multinomial logistic
    regression, 139
in ordinal logistic regression, 154
Python
packages
    `scikit-learn`, 203, 209
    `scipy.stats`, 60
    `lifelines`, 215
    `semopy`, 213
    `statsmodels`, 209
programming language, 10

R
box and whisker plots, 31

categorical factors, 14
combining dataframes, 23
data, 11
data types, 13
dataframes, 17
errors, 29
functions, 24
histograms, 31
installation and use, 10
lists, 16
loading data, 18
matrices, 16
messages, 30
missing data, 20
ordered factors, 14
packages
    brant, 157
    brglm2, 159
    broom, 204
    dplyr, 206
    frailtypack, 197
    generalhoslem, 155
    GGally, 34
    ggplot2, 33
    plotly, 34
    installing, 26
    lavaan, 174
    lme4, 168
    LogisticDx, 119
    MASS, 27, 150
    mctest, 91
    nnet, 134
    parsnip, 208
    peopleanalyticsdata, 27
    rms, 159
    survminer, 192
    survival, 189
    tidymodels, 203
    tidyverse, 29
    using, 27
    WebPower, 224
pairplot, 34
pipe operator, 28
programming language, 10
R Markdown, 34

scatter plots, 31
subsetting dataframes, 22
type coercion, 15
vectors, 14
warnings, 30
random variables, 46–49
    central limit theorem, 46
    sampling, 46
regression
    role in people analytics, 2
    binomial, *see* binomial logistic
        regression
    linear, *see* linear regression
    multinomial, *see* multinomial
        logistic regression
    ordinal, *see* proportional odds
        logistic regression
    ordinary least squares, *see* linear
        regression

Schoenfeld residual, 196
Spearman's rho, 45
standard deviation (statistics), 43
standard error (statistics), 47
statistical power, 223
stratified models, 127, 131
structural equation models, 170–184
    factor analysis, 172
    measurement model, 172, 173
    path diagram, 173
    refining, 178
    structural model, 172, 180
survival analysis, 187–199

t-distribution (statistics), 48

variance (statistics), 42
    population variance, 42
    sample variance, 42
variance inflation factor, *see* VIF
VIF (variance inflation factor), 91

Welch's t-test, 51